BTEC FIRST
Applied Science
STUDENT BOOK

John Beeby
Nicky Thomas
Lyn Nicholls
Louise Smiles
Kevin Smith

Collins

William Collins' dream of knowledge for all began with the publication of his first book in 1819. A self-educated mill worker, he not only enriched millions of lives, but also founded a flourishing publishing house. Today, staying true to this spirit, Collins books are packed with inspiration, innovation and practical expertise. They place you at the centre of a world of possibility and give you exactly what you need to explore it.

Collins. Freedom to Teach.

Published by Collins
An imprint of HarperCollinsPublishers
77 – 85 Fulham Palace Road
Hammersmith
London
W6 8JB

Browse the complete Collins catalogue at
www.collinseducation.com

10 9 8 7 6 5 4 3 2 1

ISBN-13 978 0 00 735342 2

British Library Cataloguing in Publication Data
A Catalogue record for this publication is available from the British Library

Commissioned by Letitia Luff
Project Managed by Hanneke Remsing
Edited by Jane Roth
Proofread by Elizabeth Barker
Typeset by Hedgehog Publishing
Cover design by Anna Plucinska
Production by Leonie Kellman

Printed and bound in Italy by L.E.G.O.S.p.A., Lavis TN

Acknowledgements – see page 376

This material has been endorsed by Edexcel and offers high quality support for the delivery of Edexcel qualifications. Edexcel endorsement does not mean that this material is essential to achieve any Edexcel qualification, nor does it mean that this is the only suitable material available to support any Edexcel examination and any resource lists produced by Edexcel shall include this and other appropriate texts. While this material has been through an Edexcel quality assurance process, all responsibility for the content remains with the publisher. Copies of official specifications for all Edexcel qualifications may be found on the Edexcel website - www.edexcel.com

Contents

Introduction 6

Unit 1 Chemistry and our Earth 8

Different chemical substances 10
Representing chemicals 12
Physical properties 14
Using physical properties 16
Atomic structure 18
Chemical properties of groups 1 and 7 20
Using chemical properties 24
Describing chemical reactions 26

Rates of reaction 28
Industrial processes 32
The Earth's natural activity 34
Our effect on the Earth 36
Problems we've created 38
Solutions to human problems 42
Assessment Checklist 44

Unit 2 Energy and our Universe 46

Types of energy 48
Heat transfers 50
Efficiency 52
Sources of energy 54
Environmental impact 56
Waves and communication 58
The electromagnetic spectrum 60

Ionising radiation 62
Circuits 64
Producing electricity 66
Electrical energy 68
Our solar system 70
The changing universe 72
Assessment Checklist 74

Unit 3 Biology and Our Environment 76

Cells and tissues 78
The genetic code 80
The variety of life 82
Variation and evolution 84
The interdependence of organisms 86
Human impact on the environment 88
Agriculture and the environment 92

Micro-organisms and disease 94
Carcinogens 96
The use and misuse of drugs 98
Inherited conditions 100
Diet and exercise 102
Control mechanisms 106
Assessment Checklist 108

Unit 4 Applications of chemical substances 110

Covalent bonding 112
Ionic bonding 114
Bonding and properties 116
Exothermic and endothermic reactions 118
Combustion 120
Using energy changes 122
Organic compounds 124
Alkanes and alkenes 126
Plastics 128

PVC and PVCu 130
Alcohols 132
Carboxylic acids 134
Nanostructures 136
Using nanochemistry 138
New materials 140
Smart materials 142
Assessment Checklist 144

Unit 5 Applications of physical science · 146

Measuring motion · 148
Kinetic and potential energy · 150
Forces and motion · 152
Stretching and squashing forces · 154
Effects of forces · 156
Forces on cars · 158
Parachutes and rockets · 160
Lenses · 162
Using lenses · 164
Reflection · 166
Internal reflection · 168
Sound and ultrasound · 170
Electricity and the body · 172
Electric sensors · 174
Electric cars · 176
Assessment Checklist · 178

Unit 6 Health Applications of Life Science · 180

Healthy living · 182
Food and diet · 184
Diet and health · 186
Fit and healthy · 188
Healthy heart · 190
Planning for health and fitness · 192
Preventing illness and disease · 194
The immune system · 196
Screening for disease · 198
Treatment options · 200
Antibiotics · 202
Gene therapy · 204
Assessment Checklist · 206

Unit 10 The Living Body · 208

Enzymes · 210
The digestive system · 212
The respiratory system · 216
Blood · 218
The heart and circulation · 220
Exercise and the body · 224
The kidney · 226
The nervous system · 228
Hormones · 232
Reproduction · 236
Assessment Checklist · 240

Unit 13 Investigating a crime scene · 242

Assessing the crime scene · 244
Collecting evidence · 246
Planning forensic analysis · 248
Analysing the evidence · 250
Analysing hair samples · 252
Collecting fingerprints · 254
Analysing fingerprints · 256
Testing for body fluids · 258
Testing blood samples · 260
Analysing blood patterns · 262
DNA profiling · 264
Forensic entomology · 266
Identifying remains · 268
Marks and impressions · 270
Identifying fibres · 272
Analysing glass · 274
Analysing paint · 276
Testing for drugs · 278
Identifying poisons · 280
Detecting alcohol · 282
The forensic scientist's report · 284
Forensic science and the law · 286
Assessment Checklist · 290

Unit 14 Science in medicine ### 292

Symptoms 294
Taking the temperature 296
Checking the blood pressure 298
Looking inside the body 300
Biological diagnosis 302
Microbes and disease 304
Therapeutic drugs 306
Different drugs 308
Physical therapies 310
Physiotherapy treatment 312
Surgery 314
Replacement therapies 316
Preventative therapies 318
Side-effects of drugs 320
Choice of treatment 322
The cost of treatment 324
Assessment Checklist 326

Unit 17 Chemical Analysis and Detection ### 328

Inorganic chemicals 330
Analysing cations 332
Testing for anions 334
Analysing gases 336
Acids, bases and alkalis 338
Testing for acids and alkalis 340
Uses of acids, bases and alkalis 342
Reactions of acids, bases and alkalis 344
Chromatography 346
How chromatography works 348
The chromatography of plant juices 350
Analysing unknown compounds 352
Assessment Checklist 354

Research and investigate 356

Display your results 358

Evaluate your data 360

Health and Safety 362

Glossary 364

Welcome to Collins BTEC First Applied Science!

This Student Book aims to help you achieve Pass, Merit and Distinction in the BTEC First Applied Science Certificate, Extended Certificate and Diploma qualifications.

The book is filled with exciting, relevant science content and many useful features to give you the knowledge and skills you need to complete your assignments.

The contents page shows that we have covered the core units for the Certificate, and seven other units that you can choose from to complete the Extended Certificate or the Diploma. The overview below shows you how to make the best use of your BTEC First Applied Science Student Book.

> These boxes at the beginning of each topic will help you to keep track of the assessment criteria that you have to cover in your assignments. At the end of the unit, you can use the **Assessment Checklist** to ensure that you have covered everything.

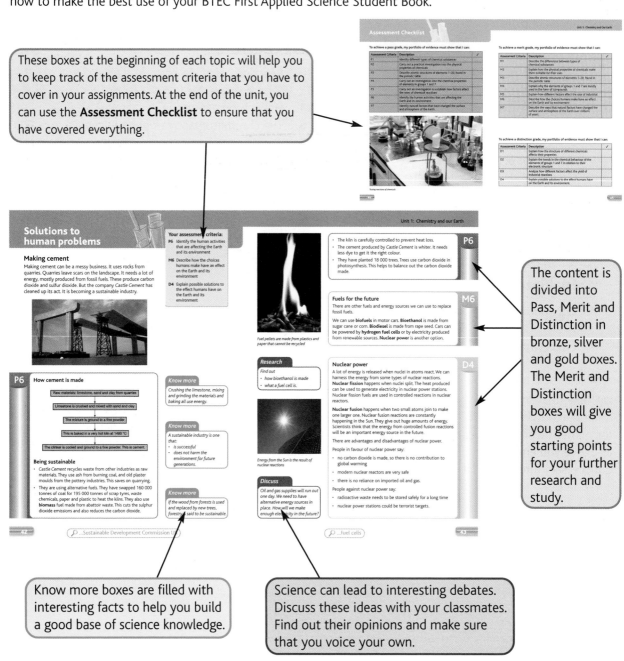

> The content is divided into Pass, Merit and Distinction in bronze, silver and gold boxes. The Merit and Distinction boxes will give you good starting points for your further research and study.

> Know more boxes are filled with interesting facts to help you build a good base of science knowledge.

> Science can lead to interesting debates. Discuss these ideas with your classmates. Find out their opinions and make sure that you voice your own.

Science is all around us. The introductions to each topic will help you put the science in a real-life context.

These 'Practical' icons show you when you need to do practical work to complete your assignments. Experiments and practical work can be dangerous. Make sure that you've read up about **Health and Safety (p. 362)** before you start.

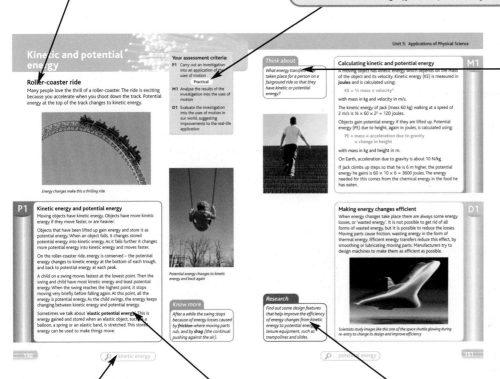

Scientists are always thinking about ideas. These boxes challenge you to think like a scientist and find solutions to many different problems.

The Internet will be a valuable resource when you do research for your assignment work. But, it can also give you endless unnecessary information if you don't search carefully. We have supplied search terms for every page to help you quickly find the right information. You can also look at the inside back cover for more Internet research tips!

Scientific terms can be intimidating. Even some everyday English words can be difficult to understand. All the bold words have been explained in the **Glossary (p. 364)** to make sure that you understand all the content.

The Research boxes give starting points to help you progress and do well in your Merit and Distinction assignments. If you are not sure where to start or how to do the research, read the sections on how to **Research and investigate (p. 356)**, how to **Display your results (p. 358)** and how to **Evaluate your data (p. 360)**.

LO

Be able to investigate the properties of elements relating to their atomic structure

- Lithium, sodium and potassium are so reactive that we have to store them under oil

- Argon gas is so unreactive that we use it to blanket metal-arc welding

- The hydrogen in heavy water has extra neutrons in its hydrogen atoms

LO

Be able to investigate different types of chemical substances related to their physical properties

- Your mobile phone contains small amounts of gold and platinum, as well as less valuable metals

- DNA and some plastic molecules are the largest molecules we know

- The periodic table contains the names and symbols of all known elements

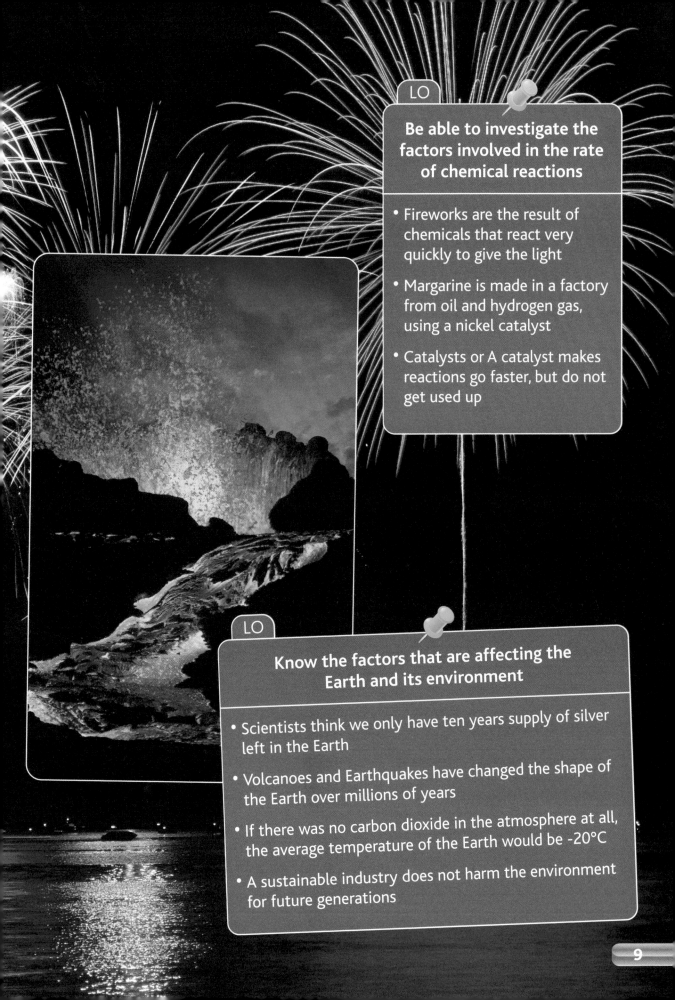

Be able to investigate the factors involved in the rate of chemical reactions

- Fireworks are the result of chemicals that react very quickly to give the light

- Margarine is made in a factory from oil and hydrogen gas, using a nickel catalyst

- Catalysts or A catalyst makes reactions go faster, but do not get used up

Know the factors that are affecting the Earth and its environment

- Scientists think we only have ten years supply of silver left in the Earth

- Volcanoes and Earthquakes have changed the shape of the Earth over millions of years

- If there was no carbon dioxide in the atmosphere at all, the average temperature of the Earth would be -20°C

- A sustainable industry does not harm the environment for future generations

Different chemical substances

Cupcakes and chemistry

Cupcakes are made from a mixture of ingredients. Different flavoured cupcakes contain different ingredients. They have different mixtures. The icing used to decorate the cakes contains sugar, water, colouring and flavouring.
Water and sugar are different types of compounds. These compounds are made from elements.

P1 Elements , compounds and mixtures

Chemical substances occur in three types.

- **Elements** – these contain one type of **atom** only. They cannot be chemically broken down into simpler substances.

- **Compounds** – these contain two or more different elements bonded together. A **chemical reaction** is needed to break up a compound. This will involve energy.

- **Mixtures** – these may contain two or more elements and/or compounds. They are mixed in any proportion and can be separated out.

When a baker mixes the flour, sugar, fat, eggs, flavouring and colour together to make cupcakes, he or she is making a mixture. The icing sugar, water and colour make a different mixture. The sugar and water are compounds.

The compound water is made from the elements hydrogen and oxygen. Sugar contains the elements hydrogen, oxygen and carbon.

Each element has a symbol. The same symbols are used in every language in the world.

Element	Symbol	Element	Symbol
hydrogen	H	copper	Cu
oxygen	O	magnesium	Mg
carbon	C	sulphur	S
iron	Fe	nitrogen	N
gold	Au	chlorine	Cl

Know more

There are 117 known elements at the moment. 94 of these are found naturally. The rest are made in nuclear reactions or in particle accelerators, like the one at CERN.

Know more

Some symbols come from the old name for an element, or from another language. Fe for iron comes from Ferrum. Au for gold comes from Aurum. Cu for copper comes from Cuprum.

Know more

Chemical substances have formulae. H_2O is the formula for water. You can tell it is a compound because the formula has two symbols.

elements...compounds...mixtures

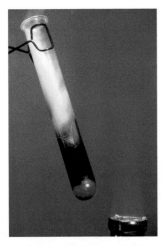

Energy is given out when iron and sulphur react

A compound from two elements

Iron is a grey shiny metallic element. It is attracted to a magnet. Sulphur is a yellow non-metallic element. It is not attracted to a magnet. They are both solid at room temperature.

When iron powder and sulphur powder are mixed together, the two elements retain their properties.

But when the mixture is heated, a chemical reaction occurs. A glow spreads through the mixture and a new substance is made. The new substance is the compound iron sulphide. It is black, non-metallic and not attracted to a magnet.

The different particles in a mixture can be in any proportion. In a compound they are always in the same proportion. In iron sulphide, the iron and the sulphur atoms are in a one-to-one ratio.

| iron atoms | sulphur atoms | iron and sulphur mixture | iron sulphide compound |

The difference between a mixture and a compound

Think about

Body lotion is a mixture of different compounds. How do chemists make sure it always has the same properties?

Looking at our coins

D1

Coins are made from special mixtures of metals called **alloys**.

1p and 2p coins, often called 'coppers', are made from copper-plated steel. Steel is a mixture of iron, carbon and other metals.

5p, 10p and 50p coins, often called 'silver', contain 75% copper and 25% nickel. £1 coins contain 70% copper, 5.5% nickel and 24.5% zinc.

Research

Why are 5p, 10p and 50p coins shiny grey but £1 coins are gold coloured, when they all contain copper and nickel? (Hint: Find out the colour of copper and nickel.)

Coins are made from different mixtures of steel, copper, nickel and zinc.

Representing chemicals

What are you made of?

Over 98% of your body is made from just six elements. Most of the atoms are joined together as molecules. About 75% of your body is water. Molecules of water are made up of hydrogen and oxygen atoms.

P1 Molecules

Two hydrogen atoms and one oxygen atom join together by **chemical bonds** to make a **molecule** of water. We write this as H_2O. This is called the formula.

The **subscript** 2 means there are two atoms of hydrogen (H) in the molecule. There is no subscript after the O, which means there is only one atom of oxygen.

We use a diagram to show how the atoms are joined together:

Methane is the main gas in natural gas. A molecule of methane contains one atom of carbon and four atoms of hydrogen. Its formula is CH_4. The diagram is:

```
      H
      |
  H — C — H
      |
      H
```

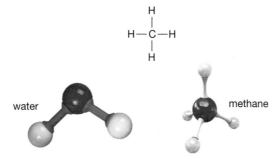

water   methane

Models of molecules show their 3D structure

Some common molecules are shown in the table.

Name	Formula
Carbon dioxide	CO_2
Hydrogen chloride	HCl
Sulphur dioxide	SO_2
Ammonia	NH_3
Glucose	$C_6H_{12}O_6$

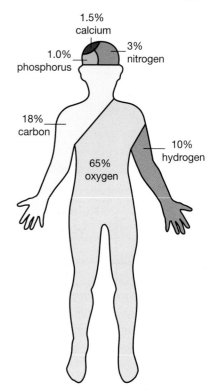

The elements in the human body

1.5% calcium
3% nitrogen
1.0% phosphorus
18% carbon
10% hydrogen
65% oxygen

Know more

Some elements occur naturally as molecules. Oxygen is O_2, chlorine is Cl_2, sulphur is S_8.

...methane

Research

Our bodies contain glucose, starch, protein and fat. Find out which elements make up these molecules.

Solids, liquids and gases

P1

Substances can exist as solids, liquids or gases. It all depends on the temperature.

- In ice, the water molecules are packed closely together. This gives the ice a definite shape and makes it hard.

- In liquid water, the molecules are close together but can move past each other. Liquids can flow and take the shape of their container.

- In steam, the molecules are very far apart. They can move anywhere. Gases can be any shape.

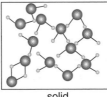 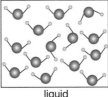 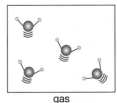

| solid | liquid | gas |

Type of atom	Number of bonds
hydrogen	1
chlorine	1
oxygen	2
sulphur	2
nitrogen	3
carbon	4

Writing formulae

M1

The table shows how many bonds some atoms make. In a molecule the number of bonds must balance. For example:

Ammonia is made from nitrogen and hydrogen.

Nitrogen makes 3 bonds and hydrogen makes 1.

The formula must be NH_3.

Think about

What is the formula of hydrogen chloride? carbon dioxide? hydrogen sulphide?

The largest molecule in your body

D1

Some molecules in our bodies are very large molecules. The largest one is **DNA**, which is found in the nuclei of our body cells. A molecule of DNA contains over 2 billion carbon atoms, and even more of other types of atoms. The order in which the atoms are bonded together in DNA is different in each of us. This unique order is our **genome** or genetic code.

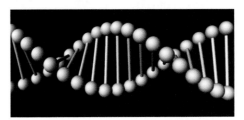

The structure of DNA was discovered in 1953

...DNA

Physical properties

Your assessment criteria:

P1 Identify different types of chemical substances

P2 Carry out a practical investigation into the physical properties of chemicals

Practical

M1 Describe the differences between types of chemical substances

M2 Explain how the physical properties of chemicals make them suitable for their uses

D1 Explain how the structure of different chemicals affects their properties

Smart mobiles

The next generation of mobile could be made from 'smart' fabric. Smart materials react to something in the environment and change. A smart fabric mobile could be folded and put in your pocket without breaking.

P1
P2

Testing physical properties

Materials must have the right properties to do their job.

Electrical conductivity

Some materials allow electricity to pass through them. These are **electrical conductors**. Other materials are **electrical insulators** and do not let electricity pass through. Metals are good electrical conductors. Non-metals usually make good insulators. In a mobile, electrical insulators surround the circuit.

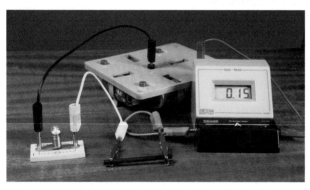

The material being tested makes up a section of the circuit. If it conducts electricity, the bulb lights up and the ammeter shows a current

Thermal conductivity

Good **thermal conductors** allow heat to pass through them efficiently. Metal kebab skewers are good thermal conductors. They take heat to the middle of the food and the kebab cooks quickly. Poor thermal conductors, such as the wooden handles of the skewers, are called **thermal insulators**.

Melting and boiling points

Each chemical substance has its own melting point and boiling point. These change if the substance has impurities. Scientists use melting and boiling points to check purity.

We use thermal insulators to keep our buildings warm

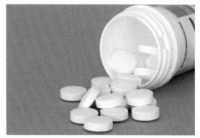

Aspirin has a melting point of 136°C. Batches of aspirin can be tested using melting point apparatus

Copper(II) sulphate is not soluble in ethanol, although it is soluble in water

'Multigrade' engine oils have polymers added. These stop the oil thinning too much when the engine is hot

Discuss

1 tonne of gold ore gives 5 g of gold metal.

1 tonne of mobile requires 150 g of gold metal.

What does this mean for our future gold supplies?

Research

Graphite is a non-metal but is a good conductor of electricity. Find out how the structure of graphite lets it conduct electricity.

Solubility

Soluble substances (sometimes called **solutes**) dissolve in a **solvent** to make a **solution**. Water and ethanol are solvents. It is important that substances used in a mobile are not soluble in everyday solvents.

Viscosity

Viscosity measures how easily a liquid flows. Thick liquids are more viscous and do not flow well. Thin liquids are less viscous.

Viscosity is an important property for engine oil. Engines need oil thin enough for a cold start and thick enough for when an engine is hot. The problem is that oil gets thinner when the engine is hot.

P1
P2

Mobile metals

Mobiles contain many different metals.

- Copper is used for electrical circuits because it is a good electrical conductor.

- Silver is used in switches on the circuit boards and in the phone buttons because it is an even better electrical conductor. It lasts for millions of on/off cycles.

- Gold is used to plate the surfaces of the circuit board and the connectors. It is an excellent electrical conductor and does not corrode.

- Tantalum is used in the electronic components. It enables scientists to make mobiles very small.

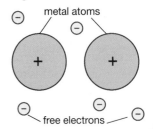

Your mobile is also likely to contain palladium, platinum, aluminium and iron.

M1
M2

How do metals conduct electricity?

Like all atoms, metal atoms have electrons orbiting the nucleus. In a solid piece of metal, some of these electrons leave their atoms and can move through the metal. These free electrons carry the electric current.

D1

Using physical properties

Your assessment criteria:

P1 Identify different types of chemical substances

P2 Carry out a practical investigation into the physical properties of chemicals

Practical

M1 Describe the differences between types of chemical substances

M2 Explain how the physical properties of chemicals make them suitable for their uses

D1 Explain how the structure of different chemicals affects their properties

Sportswear

Lycra is the trade name for spandex. It is 85% polyurethane fibres, which can stretch up to 600% of their original length. The polyurethane gives *Lycra* its stretch properties.

P1
P2

The many uses of polyurethane

Polyurethane is a man-made **plastic**. Chemists are able to make different types of polyurethane by making small changes to the chemicals they use. Different types of polyurethane have different uses. The building trade uses polyurethane insulation foam, adhesives, varnishes and glues. Polyurethane can also take the place of paint, cotton, rubber, metal or wood. It can be:

- hard like fibreglass
- squishy like a mattress
- protective like varnish
- bouncy like rubber wheels
- sticky like glue.

Making polyurethane foam

Chemical workers mix the chemicals to make polyurethane. It forms as a liquid. Liquid polyurethane is blasted out of a jet, together with blasts of carbon dioxide gas. This makes the plastic expand into a foam. The foam sets. Millions of tiny gas bubbles are trapped inside.

Either a squishy foam or a rigid foam can be produced. Squishy foam is used in sofas and mattresses. Rigid polyurethane foam is used to insulate buildings. The tiny gas bubbles are poor conductors of heat.

Polyurethane foam can be used to stop leaks

...plastics

P1 P2

Making moulded polyurethane

The liquid plastic can be poured into a mould. This can be used to make seats for airport terminals, packing crates, bumpers for cars, steering wheels and so on.

Moulded polyurethane chairs

M1 M2

Memory foam

Some mattresses and pillows are made from 'memory foam'. This is special because it moulds to the shape of your body. It is polyurethane with chemicals added to increase its viscosity and make it heavier. When you lie on it, the memory foam gets warm and moulds to the shape of your body in a few minutes. The underside stays cool and firmer to give support.

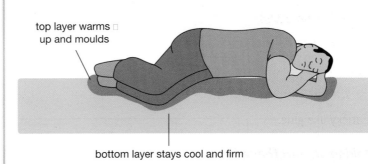

top layer warms up and moulds

bottom layer stays cool and firm

D1

Explaining polyurethane

Plastic molecules are very long. They are made from many units linked together in a long chain. Polyurethane is made from two different units. These are called **monomers**. They link together alternately. Different polyurethanes have slightly different monomers. This gives them their different properties and uses.

...polymers ...monomers

Atomic structure

Nanorobots

Nanorobots may be the technology of the future. They are microscopic machines that work on the level of atoms and molecules. Researchers are exploring many applications. They could have uses in medicine. Atomic structure is crucial in the design of nanorobots.

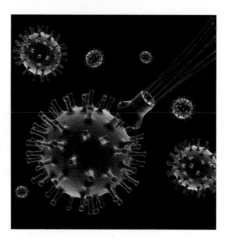

electrons, each carries a negative charge

nucleus containing protons and neutrons, carries a positive charge

The structure of an atom

P3

What's in an atom?

All atoms are made from the same basic particles. They have a central **nucleus** that contains **protons** and **neutrons**. These are collectively called **nucleons**. Around the nucleus, **electrons** move at very high speeds.

Different elements have different numbers of protons and electrons in their atoms. Each element has an **atomic number**. This is the number of protons and also the number of electrons. You can find this number on a **periodic table**.

Aluminium (Al) has atomic number 13, so an aluminium atom has 13 protons and 13 electrons.

Know more

There are even smaller particles inside protons and neutrons, called quarks. Quarks were discovered in particle accelerators.

H hydrogen 1								He helium 2
Li lithium 3	Be beryllium 4	B boron 5	C carbon 6	N nitrogen 7	O oxygen 8	F fluorine 9		Ne neon 10
Na sodium 11	Mg magnesium 12	Al aluminium 13	Si silicon 14	P phosphorus 15	S sulphur 16	Cl chlorine 17		Ar argon 18
K potassium 19	Ca calcium 20							

the atomic number of aluminium is 13

The first 20 elements of the periodic table

 ...nucleon

Know more

Hydrogen atoms with mass number 2 are called deuterium. Those with mass numbers of 3 are called tritium.

Each element also has a mass number. We can use the mass number to find out how many neutrons an atom has:

number of neutrons = mass number − atomic number

Aluminium's mass number is 27. So aluminium has 27 − 13 = 14 neutrons.

Periodic table links

Hydrogen has one proton. Helium, the next element in the periodic table, has two protons. As you go across the rows of the periodic table, each element has one more proton.

Atoms have the same numbers of electrons and protons. So the numbers of electrons also increase by one for each element.

Electrons are arranged in **shells** around the atomic nucleus. There is a limit to the number of electrons that can be in each shell.

- The first shell can have two electrons.

- The second shell can have eight electrons.

- For elements up to calcium (Ca, atomic number 20), the third shell also holds eight electrons.

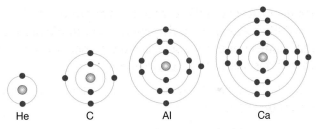

He   C   Al   Ca

Electron shells in helium, carbon, aluminium and calcium

Isotopes

Nuclear reactors use 'heavy water'. The water is heavy because the hydrogen atoms in the molecules have extra neutrons. All hydrogen atoms have one proton. Most have no neutrons. But a few hydrogen atoms have one or even two neutrons.

Atoms with the same number of protons but different numbers of neutrons are called **isotopes**. Many elements have isotopes.

All chlorine atoms have 17 protons and 17 electrons. But chlorine has two isotopes. 75% of chlorine atoms have 18 neutrons, and the other 25% have 20 neutrons.

Know more

Many heavy isotopes are radioactive because they are unstable.

Know more

Carbon has three isotopes. Carbon-12 (mass number 12) is the most abundant. The amount of carbon-14 in ancient objects is used to date them.

Chemical properties of groups 1 and 7

Halogen lights

Major supermarkets sell a variety of energy-efficient lighting. Halogen bulbs get their name from the 'halogen' elements that they contain. Halogen bulbs burn hotter and give out more light than old-fashioned light bulbs. This saves electrical energy. They also last longer.

Your assessment criteria:

P3 Describe atomic structures of elements 1–20, found in the periodic table

P4 Carry out an investigation into the chemical properties of elements in groups 1 and 7

Practical

M3 Describe the trends within the atomic structure of groups 1 and 7 in the periodic table

M4 Explain why the elements of groups 1 and 7 are mostly used in the form of compounds

D2 Explain the trends in the chemical behaviour of the elements of group 1 and 7 in relation to their electronic structure

P3 P4

About group 1

Group 1 elements are very reactive metals called the **alkali metals**. They have similar chemistry to one another, but some are more reactive than others. They react easily with water. The reactions get more violent as you go down the group.

Group 1 element	How does it react with cold water?	What's made?
lithium (Li)	fizzes	hydrogen and lithium **hydroxide**
sodium (Na)	fizzes and moves around violently	hydrogen and sodium hydroxide
potassium (K)	burns with a violet flame and moves around violently	hydrogen and potassium hydroxide

Potassium reacting with water

🔍 ... alkali metals

About group 7

A **group** in the periodic table is a column. All elements in a column have similar electron shell structures. This makes them behave chemically in a similar way.

The group 7 elements are colourful non-metals called the **halogens**. They all have similar chemical reactions, but some are more reactive than others.

They react with metals to form salts called halides. For example:

| iodine | + | sodium | → | sodium iodide |
| halogen | + | metal | → | halide salt |

The table shows the names of the halogens and their halides.

Group 7 element	Reacts with a metal to form
Fluorine (F)	a **fluoride**
Chlorine (Cl)	a **chloride**
Bromine (Br)	a **bromide**
Iodine (I)	an **iodide**

Chlorine

Bromine

PYRE

Iodine

At room temperature, chlorine is a gas, bromine is a liquid and iodine is a solid

The halogen fluorine is the most reactive element known. It bursts into flames with metals, and even attacks glass. The reactions get slower as you go down the group (down the table shown above).

P3
P4

Know more

A 'salt' in chemistry is a special type of compound.

Know more

Chlorine exists as Cl_2, bromine as Br_2 and iodine as I_2.

Two halogen atoms bond together to make a molecule.

Displacement reactions

A more reactive halogen will displace a less reactive halogen from its salt. For example, adding chlorine gas to sodium iodide solution produces iodine and sodium chloride solution.

chlorine + sodium iodide solution →
iodine + sodium chloride solution

This is called a **displacement reaction**.

Reactive elements

Group 1 and group 7 elements are not found naturally as pure elements. They are reactive elements, so have reacted with other substances to form compounds.

Sodium is a reactive element. Freshly cut sodium is shiny but it soon turns dull as it reacts with oxygen in the air

Very reactive chemicals are difficult to handle so we have few uses for them. We usually use group 1 and 7 elements in the form of compounds.

Sodium chloride is the compound made from sodium and chlorine. It is the salt we add to our food. Sodium bromide is one of the chemicals used to kill microbes in hot tubs. Sodium iodide is used as a medicine to treat iodine deficiency.

Group 1 metals are stored under oil to stop them reacting with air and water

...displacement reactions

Atomic structure in groups 1 and 7

M3

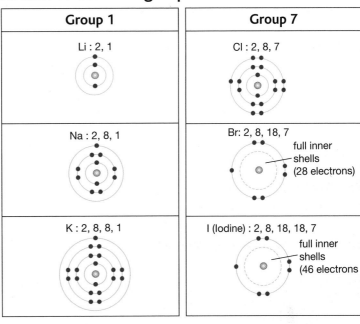

Group 1	Group 7
Li : 2, 1	Cl : 2, 8, 7
Na : 2, 8, 1	Br: 2, 8, 18, 7 — full inner shells (28 electrons)
K : 2, 8, 8, 1	I (Iodine) : 2, 8, 18, 18, 7 — full inner shells (46 electrons

You can see in the table above that all group 1 elements have one electron in the outer shell. As you go down the group, each element has an extra full shell of electrons.

All group 7 elements have seven electrons in the outer shell. They also have an extra full shell of electrons as you go down the group.

Think about

In a halogen bulb, bromine reacts with the tungsten filament. What compound is made?

Explaining the trends in reactivity

D2

Group 1

In an atom, the electrons and nucleus attract one another. When group 1 atoms react, they lose their outer electron. As you go down the group, there are more electron shells. These shield the nucleus. The outer electron is lost more easily.

So group 1 metals get more reactive as you go down the group.

Group 7

Group 7 elements gain an electron when they react. The new electron is attracted by the nucleus. Small atoms like fluorine have fewer electron shells to shield the nucleus. The nucleus can get up close and attract the new electron.

So group 7 non-metals get less reactive as you go down the group.

...uses of chlorine

Using chemical properties

Your assessment criteria:

P3 Describe atomic structures of elements 1–20, found in the periodic table

P4 Carry out an investigation into the chemical properties of elements in groups 1 and 7

Practical

D2 Explain the trends in the chemical behaviour of the elements of group 1 and 7 in relation to their electronic structure

Which shade today?

We use many chemicals every day because they 'do the right chemistry'. The dye molecules in hair dye undergo chemical reactions with your hair molecules. A chemical bond forms between them. The colour of your hair changes.

P3 P4 Some reactions involving elements 1–20

Hair dyes may be temporary or permanent. Permanent hair dyes contain many chemicals. One of these is potassium iodide. It reacts with the dyes and helps form bonds with your hair. Potassium iodide is a compound made of a group 1 metal and a group 7 non-metal.

Before digital photography became widespread, everyone used rolls of film in their cameras. The film is coated with a mixture of chemicals and has to be kept dark. When it is exposed to light, as when a photo is taken, the light starts a chemical reaction. The **products** are a different colour. This makes the image.

The chemical used on black and white film is silver bromide, which is white. When it is exposed, grey grains of silver form. This forms the darker parts of the negative. The film has to be developed to make the photo

Know more

*Silver bromide contains silver and bromine chemically bonded together. The type of bonding is called **ionic**. The silver atoms and bromine atoms become silver **ions** and bromide ions.*

...ions

P3
P4

Know more

*Silicon is a semiconductor and this is a **physical property**. Silicon reacting with oxygen is a **chemical property**.*

Silicon (atomic number 14) is used to make computer chips because of its property as a **semiconductor**. This makes it easy to control the amount of electricity flowing. Some parts of the chip, however, need to be good insulators. A chemical reaction is used. Oxygen is added to the silicon and an **oxide** is formed, silicon(IV) oxide. This compound does not conduct electricity.

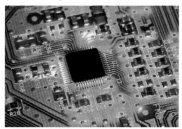

Silicon is ideal for electronic circuitry

Think about

Traditional light bulbs are filled with argon gas. Why is air not used?

Research

Find out how argon gas is obtained. Why is it cheap?

A case of unreactivity

D2

Sometimes, chemicals are used because they don't react. Arc welding is used to join metals together. It produces a lot of heat to liquefy the metals at the joint. But welders don't want the metals to burn as well. They use argon gas to blanket the welding. Argon is very unreactive (**inert**) so the metals do not burn as they would in oxygen or air.

Argon gas has been pumped over the metals to be joined, to stop them reacting

argon : 2, 8, 8

Electron shell structure of argon

Research

Which other elements have full outer shells of electrons?

Why is argon unreactive?

An argon atom has a full outer shell of electrons, as shown on the left. This makes it very stable. Atoms that do not have full outer shells of electrons undergo chemical reactions in order to become more stable. The products often have full outer shells.

...inert gas

Describing chemical reactions

An explosive reaction

Sodium metal (group 1) reacts with chlorine gas (group 7) to make sodium chloride. The reaction is spectacular. Chemistry is happening. Scientists need a way to describe what is happening.

Sodium burning in chlorine

P5 Chemical equations

Chemical equations tell us about a reaction. They use chemical symbols to represent the chemicals used and made.

A chemical equation:

- describes a real event
- tells you which chemicals are reacting, the **reactants**
- tells you which chemicals are made, the products
- uses chemical shorthand to describe the reaction
- tells you how many particles are involved.

The first step is to write a word equation:

sodium + chlorine → sodium chloride

Now use symbols and formulae for the substances:

$Na + Cl_2 → NaCl$

You must now balance the equation. List the number of each type of atom:

$Na + Cl_2 → NaCl$

    1   2   1 1

You must have the same number of each type of atom on each side of the equation. There is 1 sodium on each side, so sodium balances. There are 2 chlorines on the left-hand side and 1 on the right. Cl_2 can make two NaCl. We write this as 2NaCl.

$Na + Cl_2 → 2NaCl$

But there is now 1 sodium on the left and 2 on the right. 2Na are needed to make 2NaCl. So the **balanced equation** is:

$2Na + Cl_2 → 2NaCl$

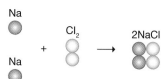

...sodium chloride

Know more

State symbols *in equations tell us more. They are added after each substance.*

(s) means solid

(l) means liquid

(g) means gas

(aq) means aqueous (dissolved in water)

Sodium reacting with water

Think about

Lithium and potassium are also in group 1. They have similar reactions with water. Can you write word and symbol equations for these reactions?

Think about

What do equations not tell us about a reaction?

P5

Remember:

- Cl_2 means there are two atoms of chlorine, chemically bonded together.

- $2Cl$ means there are two separate atoms of chlorine. They are not bonded together.

Balancing equations

M5

Sodium reacts with water to make sodium hydroxide and hydrogen gas. The gas is given off and the sodium hydroxide dissolves in the water. The word equation is:

sodium + water → sodium hydroxide + hydrogen

Using symbols:

$$Na + H_2O \rightarrow NaOH + H_2$$

Balancing:

$$Na + H_2O \rightarrow NaOH + H_2$$
$$1 \quad 2\ 1 \quad 1\ 1\ 1 \quad 2$$

There is 1 sodium on the left and 1 on the right. Sodium balances.

There are 2 hydrogens on the left and 3 on the right. $2H_2O$ will give enough hydrogen (and oxygen) to make 2NaOH. But this needs 2Na on the left. The balanced equation is:

$$2Na + 2H_2O \rightarrow 2NaOH + H_2$$

Adding state symbols:

$$2Na(s) + 2H_2O(l) \rightarrow 2NaOH(aq) + H_2(g)$$

What equations tell us

D3

Symbol equations give us more information than word equations. We can get hints about what the reaction will look like.

Calcium carbonate reacts with dilute hydrochloric acid:

$$CaCO_3(s) + 2HCl(aq) \rightarrow CaCl_2(aq) + CO_2(g) + H_2O(l)$$

An acid solution (HCl) is added to a solid ($CaCO_3$). It will probably fizz, as gas is given off (CO_2). The calcium chloride made will dissolve in the water.

...sodium hydroxide

Rates of reaction

The food additive business

Calcium chloride is a widely used food additive. It is E509. It makes food taste salty, keeps canned vegetables firm, and is added to foods to increase their calcium content. It is also used to stop the caramel in chocolate bars going hard.

P5 | Controlling the reaction

Dynamite and TNT react explosively – the reaction is very fast. The rusting of iron is a very slow reaction. We use the words **rate of reaction** to describe how fast a reaction is. It measures how much product is made in a second.

There are three ways of manufacturing calcium chloride. One way is using limestone (calcium carbonate, $CaCO_3$) and hydrochloric acid (HCl). Calcium chloride manufacturers need to control the reaction. If the reaction is too fast, there may be a dangerous explosion. If it is too slow, they will lose money.

The equation for the reaction is:

REACTANTS	PRODUCTS
calcium carbonate + hydrochloric acid	calcium chloride + carbon dioxide + water
$CaCO_3 + 2HCl$	$CaCl_2 + CO_2 + H_2O$

Manufacturers can follow the rate of reaction by two methods.

Method 1

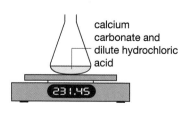

calcium carbonate and dilute hydrochloric acid

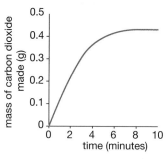

The calcium carbonate and dilute hydrochloric acid are in the flask. The mass on the balance decreases as the reaction takes place. This is because carbon dioxide gas is escaping.

Limestone, marble and chalk are all different forms of calcium carbonate. They all have the formula $CaCO_3$

⌕ ...calcium chloride food additive

Know more

As the reaction progresses, the rate of reaction decreases. This is because the reactants are being used up.

The graph shown opposite plots the mass of carbon dioxide made against time. The reaction is fastest when the graph is steepest. It is slowest when the graph is least steep. When the graph is flat, the reaction has stopped.

Method 2

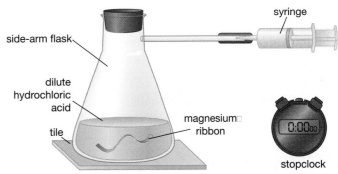

The carbon dioxide gas produced is collected in a syringe. This measures its volume. A graph can be plotted of volume of carbon dioxide against time.

Changing the rate of reaction

The manufacturers can control the reaction by:

1 changing the concentration of the reactants

2 changing the size of the reactant particles

3 changing the temperature of the reactants, or

4 using a **catalyst**.

1   Changing the concentration of the reactants

The concentration of the hydrochloric acid used can be changed. The same amount of a more concentrated acid has more acid particles to react with the limestone. The reaction will be faster. The reaction can be followed using either of the two methods described above.

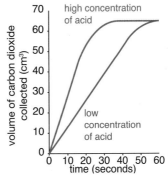

Graph to show how the rate of reaction changes when the concentration of the acid changes. You can tell that the reaction is faster with the higher concentration because the graph is steeper

2 Changing the size of the reactant particles

Limestone can have any size of particles, from large lumps to fine powder. The smaller the particles, the faster the reaction with hydrochloric acid. Powdered limestone reacts fastest.

3 Changing the temperature of the reactants

Consider a different reaction: sodium thiosulphate reacting with dilute hydrochloric acid. Small particles of sulphur form. These make the mixture go cloudy. You can tell how long the reaction takes by timing how long it takes for a cross under the flask to disappear.

This reaction can be carried out at different temperatures. Most reactions are faster at higher temperatures.

4 Using a catalyst

A catalyst is a substance added to a chemical reaction to make it go faster. It is not used up in the chemical reaction.

Manufacturers find catalysts very useful. They can make the product more quickly and use the catalyst over and over again.

Consider the reaction of hydrogen peroxide, H_2O_2. Left on its own, it slowly breaks down into oxygen gas and water. A catalyst will make this happen quicker.

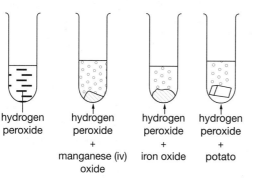

| hydrogen peroxide | hydrogen peroxide + manganese (iv) oxide | hydrogen peroxide + iron oxide | hydrogen peroxide + potato |

Know more

1.0 g of powdered limestone has a larger total surface area than a 1.0 g lump of limestone. There are more places for the reaction to happen and so the reaction is faster.

Catalytic converter

Know more

Catalytic converters use platinum and other precious metals as catalysts. They change harmful exhaust gases into carbon dioxide and nitrogen.

...precious metal

Before	After

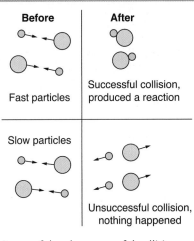

Fast particles	Successful collision, produced a reaction
Slow particles	Unsuccessful collision, nothing happened

Successful and unsuccessful collisions

The collision theory

Manufacturers know that a reaction will only take place when particles collide. They must be moving very fast – have lots of **kinetic energy**. Only particles with enough energy will react. These are the **successful collisions**.

The **collision theory** explains why changes in concentration, particle size and temperature change the rate of reaction.

Change made	Explanation
concentration	low concentration high concentration ⬤ Reacting particle of substance A ⬤ Reacting particle of substance B At high concentration, the particles are more crowded. Collisions are more frequent. The rate of reaction increases.
particle size	When substance B is broken into small pieces, there are more places for collisions to happen. The rate of reaction increases.
temperature	low temperature high temperature ⬤ Reacting particle of substance A ⬤ Reacting particle of substance B Particles move faster at high temperatures. Collisions are more frequent. The rate of reaction increases.

Think about

If colliding particles do not have enough energy to react, what happens to them?

Research

Research other industrial reactions that use catalysts.

...kinetic energy

Industrial processes

Green chemistry

Chemical manufacturers must make money. Managers must consider the rates of reaction and the yield of products. These days they must also use 'green chemistry'. Green chemistry produces as little environmental harm as possible. Ibuprofen is now made in three steps instead of six. There is less waste and less pollution.

Green Chemistry

Centre of Excellence

P5

Making natural materials more useful

Making margarine

Fats are solid at room temperature, whereas oils are liquid. Bakers often prefer solid cooking fats like margarine for baking. Food manufacturers can make margarine and cooking fats from plant oils. If hydrogen gas is bubbled through oil for a long time, a hard fat is made. The method is too slow to make a profit. Food manufacturers speed up the reaction by:

- using a nickel catalyst
- using a high pressure – this concentrates the hydrogen gas.

Making synthetic rubber

Rubber can be natural or synthetic (man-made). Natural rubber comes from the latex in rubber trees. Scientists make synthetic rubber from oil and natural gas. About 70% of all rubber used is synthetic.

Latex is collected from a rubber tree

Making fertiliser

Some fertilisers contain ammonia. Ammonia is NH_3. It is made in a chemical plant (factory) from nitrogen and hydrogen. Nitrogen comes from the air and hydrogen from natural gas. Ammonia is made by the **Haber process**.

> **Know more**
>
> *In 1870 Napoleon challenged scientists to make a cheap butter substitute that he could feed to his troops. Margarine was discovered.*

> **Know more**
>
> *Oils such as olive oil are **unsaturated fats**. Margarine is a **saturated fat**. Saturated fats have health risks for your heart and blood vessels.*

$\mathcal{P}$ ...green chemistry principles

P5

Plants use the nitrogen in ammonia-based fertiliser to make protein

The Haber process

The reaction equation is:

nitrogen + hydrogen $\rightleftharpoons$ ammonia

$$N_2 + 3H_2 \rightleftharpoons 2NH_3$$

In the equation, $\rightleftharpoons$ means that it's a **reversible reaction**.

Nitrogen in the air is not very reactive. Chemists have to control the reaction to get any ammonia at all. They make the reaction happen by:

- increasing the pressure – this makes the gases more concentrated, pushing the molecules closer together
- using a temperature of 400 °C to speed up the reaction
- using a iron catalyst.

D3

The yield of products

Atom economy is an important idea in green chemistry. It measures how much of the reactants ends up in the useful product, and how much ends up as waste. Some chemical wastes are harmful to the environment. Wasteful processes have low atom economies. Efficient processes have high atom economies.

The old methods used to make ibuprofen had an atom economy of 40%. Using green chemistry has improved the atom economy to 80%, and 99% of the catalyst is recycled.

When chemical plant workers measure out the reactants, they know from the atom economy how much product they should be able to make. This would be 100% **yield**. In practice this rarely happens. The yield for making ammonia is just 10 to 20%.

$$\text{Percentage yield} = \frac{\text{actual mass of product obtained}}{\text{calculated mass of product}} \times 100\%$$

Not all atoms do the planned chemistry. In some processes, there are side reactions making different unwanted products. Also, some reactions can go both ways – they are reversible. As nitrogen reacts with hydrogen to make ammonia, ammonia also breaks down to make nitrogen and hydrogen. Manufacturers choose the best conditions of temperature, concentration, particle size and catalyst to give the highest yield of the desired product, as quickly as possible.

Think about

Atom economies are calculated from the equation. This involves working out the percentage of the mass of the product out of the mass of all the reactants. What is the atom economy for making ammonia?

The Earth's natural activity

Your assessment criteria:

P7 Identify natural factors that have changed the surface and atmosphere of the Earth

M7 Describe the ways that natural factors have changed the surface and atmosphere of the Earth over millions of years

The Earth is moving

Mount Etna is in Italy. It is the largest active volcano in Europe. It is always active. Mount Etna regularly spews out molten rock or lava, hundreds of metres into the air. Its lava flows threaten and sometimes cover small villages. But the volcanic ash makes rich soil. People take a chance living near the volcano to farm the fertile soil.

P7 Why do we have volcanoes?

We live on the thin rocky **crust** of the Earth. Beneath us is the **mantle** and at the centre an **iron core**. The mantle is semi-liquid and moves slowly.

The Earth's crust is made up of **tectonic plates**. They float on the mantle and move very slowly. At plate boundaries the movements of the plates cause **earthquakes** and **volcanoes**. The map shows where they occur.

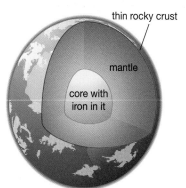

The Earth's layers

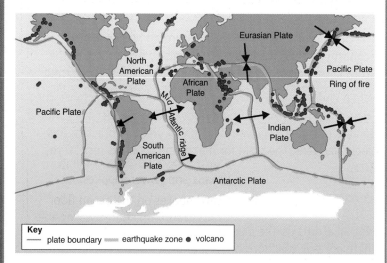

Key
— plate boundary ▬ earthquake zone ● volcano

Tectonic plate boundaries

What happens in a volcano?

Molten rock (**magma**) can work its way through weak spots in the Earth's crust. When it reaches the surface, it erupts as **lava**. This then cools and solidifies. Layers of lava build up and make the volcano mountain. Volcanoes can erupt many times.

Know more

The Earth's crust is between 10 and 50 km thick. The thinnest parts are under the oceans.

🔍 ...tsunami

As well as lava, volcanoes produce volcanic gases. These may cause explosions. The common gases given off are water vapour, carbon dioxide and sulphur dioxide. Carbon dioxide and sulphur dioxide are heavy gases. They can form a suffocating blanket over living things. Chlorine and fluorine can also be produced.

Inside a volcano

How do earthquakes happen?

Tectonic plates move slowly. Some move apart, others move together, some slide past each other. If the plates get stuck, pressure builds up. An earthquake occurs when the plates move again.

Know more

The city of Pompeii was buried in 20 m of volcanic ash and rock when Mount Vesuvius erupted in AD79. Few survived.

The ground literally moves during an earthquake

Volcanic effects

Volcanoes and earthquakes change the environment.

Good changes	Bad changes
Volcanic ash and lava make very fertile soil.	Lava flows destroy crops.
Volcanoes on the ocean floor eventually make new islands.	Volcanic gases are harmful. Sulphur dioxide makes acid rain. Carbon dioxide is a greenhouse gas. These are both heavy gases that can suffocate living things. Chlorine and fluorine are poisonous.
Volcanic mountains can give rise to beautiful landscapes.	Gases dissolve in rivers, lakes and oceans. This makes the water more acidic. Wildlife can be harmed.
	Earthquakes destroy buildings, roads and power supplies.
	Earthquakes on the ocean floor cause **tsunamis** (large waves able to destroy coastal areas).

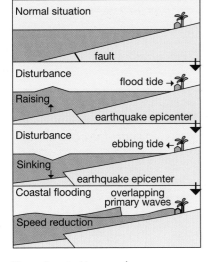

How a tsunami is caused

Early warnings

Geologists study volcanoes and tectonic plate movements to try to predict activity. They use **seismographs** to detect vibrations in the Earth. Satellites look for changes in the shape of volcanoes, which may mean that magma is getting nearer the surface.

Research

Find out how Iceland was made from volcanoes on the ocean floor.

Our effect on the Earth

Your assessment criteria:

P6 Identify the human activities that are affecting the Earth and its environment

M6 Describe how the choices humans make have an effect on the Earth and its environment

D4 Explain possible solutions to the effect humans have on the Earth and its environment

It's not rocket science

But we may need the rocket science to get materials from the Moon. We are using up Earth's materials faster than ever before. We still have large deposits of iron and titanium but other metals are running out. Indium is used to make flat-screen televisions. We have about ten years' supply left. Silver will also run out in ten years. Is it worth developing new technologies if we haven't got the metals needed?

P6 Materials from the Earth

From the sea

Sea water contains dissolved salts. We evaporate it to get sea salt. Salt ponds can be found along the coasts of warmer countries. We also process sea water to take out the bromine. Bromine is used in many medicines and in flame-proofing.

From the land

We get our **fossil fuels** from the land. **Coal** deposits are found underground and on the surface. Coal formed naturally 300 million years ago. It is fossilised plant material.

Oil and **natural gas** are found together. They are the remains of plants and animals that lived in the oceans 100 million years ago.

Few metals are found naturally as elements. Most are in compounds. Iron is found as iron oxide. This iron **ore** is mined and the iron is extracted in a **blast furnace**. We take millions of tonnes of metal ore from the Earth every year.

Quarries are places where ore or building stone is dug out. All of our buildings and roads are made from different types of rock. St Paul's Cathedral was made from Portland limestone. The tarmac on our roads is a mixture of crushed rock and tar (from oil).

Know more

*Fossil fuels are finite, or **non-renewable**. There is only so much in the Earth. Once we have used that up, we can't replace it.*

Oil is extracted from deep beneath the ocean floor

Know more

Gold, silver and platinum are found as elements in the Earth's crust.

 ...coal formation

P6

Research

Find out how iron is extracted from iron ore. What are the waste materials and what happens to them?

From the air

We have many uses for the gases in air (see the table). We separate them out from the mixture of gases using **fractional distillation**. Air is cooled until it becomes liquid.

As it slowly warms up, gases boil off at different temperatures.

Gas	How do we use it?
oxygen	Breathing apparatus, welding.
nitrogen	Making fertilisers, explosives, modified atmosphere packaging of food.
argon	In light bulbs, plasma globes, blue laser lights, arc welding.
helium	Diving gas (mixed with oxygen for divers to breathe), filling balloons, helium-neon lasers.

M6

Chemical processing

Most of the substances we take from the Earth have to be processed to make them useful. This can involve **physical changes** and **chemical changes**, which need energy. We obtain most of our energy from fossil fuels. When we burn these, carbon dioxide and pollutants are given off.

Metal extraction processes often have low atom economies. This means a lot of waste is produced. This may be piled in slag heaps, changing the landscape. It can be **toxic**.

Think about

How will green chemistry help reduce slag heaps?

Waste heaps are unpleasant to look at and can be toxic

D4

Think about

Platinum is a very expensive metal. It is used in catalytic converters. Small amounts may be lost in car exhaust. Should we recycle dust from road sweepers?

Solutions

In the UK we have local planning regulations and government bodies to control material extraction and building. Mining licences have to be obtained before companies can start digging. Planning permission must be obtained to put up new buildings. The Environment Agency keeps a close watch on possible environmental impacts.

🔍 ...oil formation

Problems we've created

The Thames barrier

The Thames barrier protects central London from flooding due to high tides. Its floodgates have been used increasingly often since 1990. This is because the sea level is rising by 3 mm per year. Global warming is blamed for this rise. The sea level is expected to rise one metre by 2100. Plans are underway to improve the barrier to cope with this.

P6 P7 The problems

Global warming

Scientists agree that the Earth is getting warmer. Most think this is because we are burning large amounts of fossil fuels. Fossil fuels contain carbon. When they burn, carbon dioxide is made. Carbon dioxide is a **greenhouse gas**. It absorbs heat energy in the atmosphere. The Earth gets hotter.

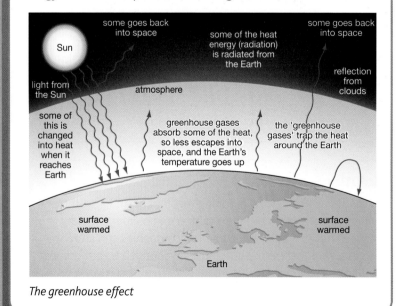

The greenhouse effect

Know more

There are several greenhouse gases. Water vapour and methane are also greenhouse gases. We need some greenhouse gas in the atmosphere. Without carbon dioxide, the average temperature of the Earth would be −20 °C.

🔍 ... greenhouse gases

Ozone issues

There is a layer of **ozone** gas high in the atmosphere. This absorbs some of the harmful UV radiation from the Sun. UV radiation can cause skin cancer. Chemicals called **CFCs** used to be used in aerosols and refrigerators. These break down the ozone high in the atmosphere. When ozone is destroyed more harmful UV radiation can reach the Earth.

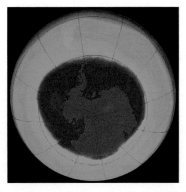

There are now holes in the ozone layer over the South Pole; ozone layer thickness is colour coded from purple (lowest) to green (highest)

Acid rain

Many fuels contain sulphur impurities. The sulphur makes sulphur dioxide when the fuel burns. Sulphur dioxide in the air reacts with water and oxygen producing sulphuric acid. This makes rain water acidic.

Oxides of nitrogen are made in car engines and escape in the exhaust. These also make **acid rain**.

Acid rain changes the chemistry of lakes and rivers. Fish are killed. It damages trees, other plants and buildings.

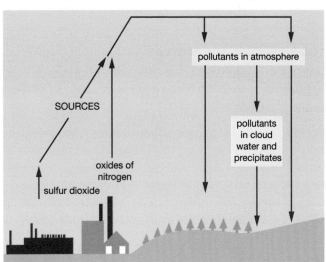

How acid rain is formed

> ### Know more
> Most countries now use different chemicals (not CFCs) in aerosols and refrigerators. But some of these replacement chemicals act as greenhouse gases.

> ### Know more
> Unpolluted rain is pH 5 to 6. Acid rain is pH 2 to 4.

> ### Know more
> Power stations making electricity from fossil fuels have been the major sulfur dioxide producers.

...CFCs

Adverse effects of chemicals

An inherited problem – land contamination

In some areas of the UK, 200 years of industry has left many chemicals in the soil. Some of these **contaminants** are hazardous. Many sites such as old chemical factories, disused mines and places where waste was tipped are now classed as contaminated sites. The Government now requires developers to clean up land before they build on it.

A *current problem – sheep dips*

Sheep farmers regularly wash their sheep in an insecticide solution called sheep dip. The dip kills the parasites in the wool. But the used dip can end up in rivers and streams, where it kills invertebrates and fish. Sheep dip chemicals also cause health problems for some users. Organic farmers are developing alternative methods to deal with parasites.

A success story

Acid rain was discovered in the 1960s. Since then:

- many governments have made energy producers clean up the smoke from their tall chimneys

- money has been provided to restore ecosystems damaged by acid rain

- use of clean, **renewable energy** has been encouraged.

Sulphur dioxide emissions from industry in the UK have now fallen by 80%. Environmentalists say our wildlife is recovering.

Research

What are the alternatives to using sheep dip?

Think about

How does using renewable energy sources to produce electricity help reduce sulphur dioxide in the air?

Discuss

Is the problem of acid rain over? Can we now write about it in the history books?

Carbon capture and storage

'Carbon capture and storage', or CCS, does what it says. Firstly, it captures the carbon. Secondly, it stores it safely so that it can't act as a greenhouse gas.

The carbon in a fuel produces carbon dioxide when the fuel burns. CCS involves capturing this carbon dioxide before it gets to the atmosphere. Old oil and gas fields are good places for storing the carbon dioxide. They are large enough to store millions of tonnes. The carbon dioxide is injected underground. With less carbon entering the air as carbon dioxide, global warming is reduced.

Think about

Capturing carbon dioxide from power stations is possible. But what about car exhausts?

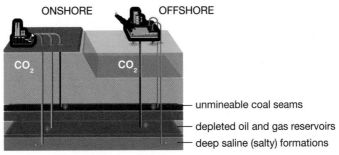

ONSHORE   OFFSHORE

CO_2 CO_2

— unmineable coal seams
— depleted oil and gas reservoirs
— deep saline (salty) formations

Storing carbon dioxide (CO_2) underground

Renewable energy

Renewable energy sources are naturally replenished. We can get energy from sunlight, wind and moving water, and we can get geothermal heat from the Earth. Using renewable energy does not produce carbon dioxide. Global warming is reduced.

Solar energy

Energy from the Sun can be used in different ways:

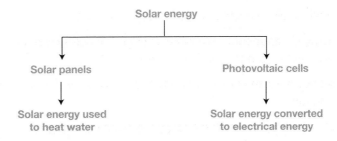

Solar energy

Solar panels → Solar energy used to heat water

Photovoltaic cells → Solar energy converted to electrical energy

Wind turbines are non-polluting

Wind energy

Energy from the wind is used to turn the blades on a wind turbine. These are connected to a generator which produces electricity.

Energy from water and tides

Water turbines work in a similar way to wind turbines. Fast-moving streams of water, or the changing sea level as the tide ebbs and flows, are used to turn paddles that are connected to an electricity generator. We have very high tidal ranges in the UK. When technical problems have been overcome, tidal turbines just offshore could provide a major part of the UK's electricity.

Geothermal energy

The deeper you go inside the Earth, the hotter it gets. **Geothermal energy** plants use some of this heat energy to heat water and make steam. The steam can drive turbines and generate electricity.

...solar energy uses

Solutions to human problems

Making cement

Making cement can be a messy business. It uses rocks from quarries. Quarries leave scars on the landscape. It needs a lot of energy, mostly produced from fossil fuels. These produce carbon dioxide and sulfur dioxide. But the company *Castle Cement* has cleaned up its act. It is becoming a sustainable industry.

Your assessment criteria:

P6 Identify the human activities that are affecting the Earth and its environment

M6 Describe how the choices humans make have an effect on the Earth and its environment

D4 Explain possible solutions to the effect humans have on the Earth and its environment

P6

How cement is made

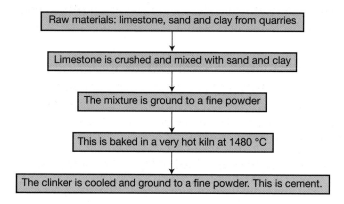

Raw materials: limestone, sand and clay from quarries

↓

Limestone is crushed and mixed with sand and clay

↓

The mixture is ground to a fine powder

↓

This is baked in a very hot kiln at 1480 °C

↓

The clinker is cooled and ground to a fine powder. This is cement.

Being sustainable

- *Castle Cement* recycles waste from other industries as raw materials. They use ash from burning coal, and old plaster moulds from the pottery industries. This saves on quarrying.
- They are using alternative fuels. They have swapped 160 000 tonnes of coal for 195 000 tonnes of scrap tyres, waste chemicals, paper and plastic to heat the kilns. They also use **biomass** fuel made from abattoir waste. This cuts the sulphur dioxide emissions and also reduces the carbon dioxide.

Know more

Crushing the limestone, mixing and grinding the materials and baking all use energy.

Know more

A sustainable industry is one that:
- *is successful*
- *does not harm the environment for future generations.*

Know more

If the wood from forests is used and replaced by new trees, forestry is said to be sustainable.

...Sustainable Development Commission UK

Fuel pellets are made from plastics and paper that cannot be recycled

P6

- The kiln is carefully controlled to prevent heat loss.
- The cement produced by *Castle Cement* is whiter. It needs less dye to get it the right colour.
- They have planted 18 000 trees. Trees use carbon dioxide in photosynthesis. This helps to balance out the carbon dioxide made.

M6

Fuels for the future

There are other fuels and energy sources we can use to replace fossil fuels.

We can use **biofuels** in motor cars. **Bioethanol** is made from sugar cane or corn. **Biodiesel** is made from rape seed. Cars can be powered by **hydrogen fuel cells** or by electricity produced from renewable sources. **Nuclear power** is another option.

Research

Find out
- *how bioethanol is made*
- *what a fuel cell is.*

Energy from the Sun is the result of nuclear reactions

Discuss

Oil and gas supplies will run out one day. We need to have alternative energy sources in place. How will we make enough electricity in the future?

D4

Nuclear power

A lot of energy is released when nuclei in atoms react. We can harness the energy from some types of nuclear reactions. **Nuclear fission** happens when nuclei split. The heat produced can be used to generate electricity in nuclear power stations. Nuclear fission fuels are used in controlled reactions in nuclear reactors.

Nuclear fusion happens when two small atoms join to make one larger one. Nuclear fusion reactions are constantly happening in the Sun. They give out huge amounts of energy. Scientists think that the energy from controlled fusion reactions will be an important energy source in the future.

There are advantages and disadvantages of nuclear power.

People in favour of nuclear power say:

- no carbon dioxide is made, so there is no contribution to global warming
- modern nuclear reactors are very safe
- there is no reliance on imported oil and gas.

People against nuclear power say:

- radioactive waste needs to be stored safely for a long time
- nuclear power stations could be terrorist targets.

...fuel cells

Assessment Checklist

To achieve a pass grade, my portfolio of evidence must show that I can:

Assessment Criteria	Description	✓
P1	Identify different types of chemical substances	
P2	Carry out a practical investigation into the physical properties of chemicals	
P3	Describe atomic structures of elements 1–20, found in the periodic table	
P4	Carry out an investigation into the chemical properties of elements in groups 1 and 7	
P5	Carry out an investigation to establish how factors affect the rates of chemical reactions	
P6	Identify the human activities that are affecting the Earth and its environment	
P7	Identify natural factors that have changed the surface and atmosphere of the Earth.	

Testing reactions of chemicals

To achieve a merit grade, my portfolio of evidence must show that I can:

Assessment Criteria	Description	✓
M1	Describe the differences between types of chemical substances	
M2	Explain how the physical properties of chemicals make them suitable for their uses	
M3	Describe atomic structures of elements 1–20, found in the periodic table	
M4	Explain why the elements of groups 1 and 7 are mostly used in the form of compounds	
M5	Explain how different factors affect the rate of industrial	
M6	Describe how the choices humans make have an effect on the Earth and its environment	
M7	Describe the ways that natural factors have changed the surface and atmosphere of the Earth over millions of years.	

To achieve a distinction grade, my portfolio of evidence must show that I can:

Assessment Criteria	Description	✓
D1	Explain how the structure of different chemicals affects their properties	
D2	Explain the trends in the chemical behaviour of the elements of groups 1 and 7 in relation to their electronic structure	
D3	Analyse how different factors affect the yield of industrial reactions	
D4	Explain possible solutions to the effect humans have on the Earth and its environment.	

Unit 2 Energy and our Universe

Be able to investigate energy transformations

- Since the Earth was formed, the only source of additional energy is the Sun; the energy you use has been recycled from other forms in the past

- Sources of renewable energy include cow dung, and peat cut from bogs and dried

- Many refrigerated lorries are painted light colours; this reduces energy absorbed by thermal radiation so the energy needed to keep lorries cool is reduced

- The Three Gorges Dam in China will be the world's largest hydroelectricity scheme; its reservoirs will destroy towns, forests, habitats of native animals and disrupt the flow of water above and below the dam; however, flooding from the river which has killed thousands in previous years will be easier to control

LO

Know the properties and applications of waves and radiation

- The first radio broadcast of a voice was in 1906; a message was sent from Boston, America and received in Machrihanish, Scotland

- Different animals see different frequencies of the electromagnetic spectrum; humans see visible light; some snakes, fish and birds "see" using infrared radiation; some fish, birds and insects like bees can see using ultraviolet radiation

- Our biggest exposure to ionising radiation is from our surroundings; people living in regions where granite is found are likely to have higher rates of lung cancer due to radioactive radon gas emitted from the rock

Know how electrical energy that is produced... can be transferred...

- Pacemakers contain tiny electrical circuits that send a signal to the heart to keep it beating in the correct rhythm

- Abergeirw in Wales was the last village in the UK to get a supply of mains electricity in 2008; before then they used noisy diesel generators that could not be left on all the time

- During the day there are surges in demand for electricity; people controlling the National Grid anticipate these surges and shut down or startup power stations to match the demand with the supply

Know the components of the solar system and the way the Universe is changing

- Our Solar System has 8 planets and 4 dwarf planets – Pluto, Ceres, Eris and Makemake

- Most of the mass of the Solar System is the Sun – 99.86%

- The future of the Universe depends on how much matter is in it; too little matter and it will keep expanding, or too much and it will start to contract again and we will have the Big Crunch

Types of energy

Energy and sport

You're at a large sporting event. The spectators jump to their feet as the players run and shout, and the sound system blasts out a loud announcement.
Lots of different types of energy are involved.

Your assessment criteria:

P1 Carry out practical investigations that demonstrate how various types of energy can be transformed

Practical

M1 Describe the energy transformations and the efficiency of the transformation process in these investigations

D1 Explain how energy losses due to energy transformations in the home or workplace can be minimised to reduce the impact on the environment

P1 Types of energy

The main types of energy are listed here.

- **Thermal energy** – everything has thermal energy. Hotter objects have more thermal energy.

- **Light energy** – anything glowing, such as light bulbs, gives out light energy.

- **Sound energy** – anything making a noise gives out sound energy.

- **Chemical energy** – found in food and fuels. Fuels give out energy when burned.

- **Electrical energy** – equipment that uses a plug or batteries uses electrical energy.

- **Kinetic energy** – moving objects have kinetic energy.

- **Potential energy** – another word for stored energy. Energy is stored when things are lifted up or when stretchy objects are stretched or squashed.

- **Nuclear energy** – this is stored inside atoms. Inside the Sun, nuclear energy is changed into thermal and light energy.

Whenever anything happens, energy changes take place. These can be **energy transformations** (when the energy changes from one type to another), or **energy transfers** (when the energy moves from one object to another, or one region to another).

- A car engine transforms chemical energy into kinetic energy.

- A kettle transfers thermal energy from its heating element to the water.

Know more

Energy is measured in **joules** (J). You use about one joule of energy picking up a football.

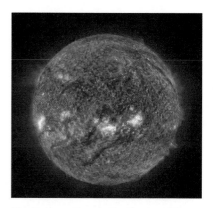

Nuclear reactions in the Sun provide all our energy on Earth

Know more

Several types of energy can be involved. An exploding firework gives out heat, light and sound, as well as particles with kinetic and potential energy.

...types of energy

Many different energy changes take place in a tennis match

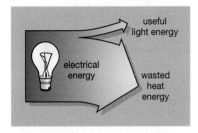

Block diagrams are helpful in showing energy changes

Think about

List all the energy changes that take place on your journey to school.

Discuss

Which forms of transport shown here waste the least energy?

Energy transformations

M1

Energy changes can take place in more than one stage.

- Chemical energy in food eaten by a tennis player is transformed into kinetic energy when she serves. Kinetic energy is transferred from the player to the racket, then to the tennis ball.

More than one main energy transformation can take place.

- An electric bulb transforms electrical energy into light energy and thermal energy.

- A climber transforms chemical energy into kinetic energy, and potential energy.

- A solar powered fan changes light to electrical energy, which is changed into mechanical energy.

- A battery powered electric drill changes chemical energy to electrical energy and then into mechanical energy.

- A nuclear power station changes nuclear energy to thermal energy, then mechanical energy and then electrical energy.

Energy losses

D1

If the energy changes into types that we can't easily use, such as sound or thermal energy, it is called wasted energy. Some energy is wasted at each energy change. Machines that cause energy changes in several stages waste a lot of energy.

- A light bulb gives out light energy and unwanted thermal energy because it heats up as it glows.

- A car engine causes wasted sound energy and thermal energy (as the wheels and engine parts warm up).

...energy transformation

Controlling the heat

Keeping buildings at the right temperature is difficult when lots of people are coming and going.

Both buildings are designed to control heat transfers; the materials used, the shapes of doors and windows, and the size of rooms all have an effect

P1 Types of heat transfer

Thermal energy (or heat) always moves from hot places to cooler places. Buildings lose heat in winter because it is cooler outside. The three ways that heat is transferred are **conduction**, **convection** and **radiation**.

Conduction

- If one part of a solid object is heated, particles inside the solid vibrate more. The vibration passes thermal energy to nearby particles which vibrate more, and so on.

- Some materials, especially metals, transfer thermal energy quickly. They are good heat **conductors**.

- Other materials transfer thermal energy only very slowly. Building materials such as wood and uPVC slow down heat losses. These are **insulators**.

Convection

- Heat spreads through a gas or liquid when particles in the warmer region gain thermal energy and move into the cooler regions carrying this energy with them.

- These **convection currents** spread heat quickly. Warm air above a radiator rises and cool air sinks to replace it. As the air circulates, thermal energy spreads around the room.

- Putting a lid on a hot drink slows down convection in the air above the drink, so it stays hot for longer.

The car heats up quickly because metal and glass are good heat conductors

Convection makes smoke rise from a log fire

...conduction ...convection ...radiation

White buildings reflect thermal radiation and stay cool on hot sunny days

Thermal radiation

- Something hotter than its surroundings radiates heat from its surface as waves of thermal radiation. Hot surfaces radiate more heat than cooler surfaces.

- Something colder than its surroundings will absorb thermal radiation, and warm up.

- Black, dull surfaces are good at absorbing and at giving out thermal radiation. They warm up and cool down quickly.

- Light shiny surfaces reflect thermal radiation and so slow down heat transfer.

Think about

What has been done to reduce unwanted heat transfers in the room you are in?

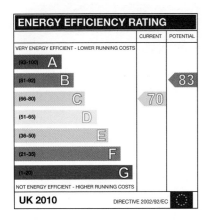

This shows the efficiency rating of a house now and how energy-efficient it could be if improvements were made

Think about

Suggest what measures might be taken to increase the efficiency rating of a house.

Making heat transfers work for us

We can design containers and buildings to slow down heat transfers.

- A drink stays hot for longer in a white polystyrene cup with a lid.

- Many homes have insulation in the walls and loft, as well as having carpets, curtains and double glazing to reduce heat transfers.

The cup slows down heat transfers

An energy transformation or transfer is said to be **efficient** if unwanted heat losses are minimised.

Sometimes, we want to speed up heat transfers.

- A saucepan made of metal with a black coating will cook food quickly.

- Computers need to stay cool to work well. Cooling systems include 'heat sinks' made of metal painted black to radiate heat, and fans to circulate air.

Saving the environment

Well insulated buildings with efficient electrical appliances reduce energy losses. This saves money because less energy is wasted, and also saves the environment because fewer resources are used to heat the building or produce electricity. If you sell a home, it must have an energy rating based on its structure, its heating equipment and other appliances.

...heat transfer

Efficiency

Your assessment criteria:

P1 Carry out practical investigations that demonstrate how various types of energy can be transformed

Practical

P2 Calculate the efficiency of energy transformations

M1 Describe the energy transformations and the efficiency of the transformation process in these investigations

D1 Explain how energy losses due to energy transformations in the home or workplace can be minimised to reduce the impact on the environment

Science can help you win

Sports matches and races can be very exciting and usually the best participants win. If their equipment isn't in the best condition though, they waste energy that should be used to succeed in their sport. Top sportspeople make sure their equipment is in good condition so they can perform efficiently and save their energy for winning.

P1 What does 'efficient' mean?

All cyclists change chemical energy from their food into useful kinetic energy, as well as wasted thermal energy and sound energy. If a bicycle is specially designed to be efficient, the same person will cycle faster, especially if the bike is looked after well. This is because less chemical energy is changed into wasteful forms of energy. An efficient bike will help a cyclist win a race because more of the cyclist's energy is used to make it travel fast.

Toe clips make bikes more efficient because the pedals are pulled up as well as pushed down

P2 Calculating efficiency

The **efficiency** of an energy change can be calculated:

efficiency = useful energy output ÷ total energy input

This means an energy change is more efficient if:

- less energy is wasted
- more energy is usefully changed.

Efficiency is always less than 1 (or 100%) because some energy is always wasted in any energy change. The efficiency of electrical equipment can be very high, but some energy is still wasted in heating up connections inside the equipment.

Know more

A cyclist's efficiency is only about 0.2, or 20%. This means that only 20%, or a fifth, of the energy from the food eaten by the cyclist is used to make the bike move.

...calculating efficiency

The efficiency of this bulb is 0.04 (4%); almost all of the electrical energy is wasted as heat

P2

The right sports equipment improves efficiency.

- Strings on a tennis racket are tight, so energy is transferred efficiently from the racket to the ball.
- A football is inflated hard, so energy is transferred efficiently when it is kicked.

The right sports clothes also improve the efficiency of the sportsperson.

- Cyclists and swimmers wear close-fitting clothes to reduce drag forces.
- Runners wear lightweight shoes.

Think about

Some light bulbs are more efficient than others. An efficient bulb produces more light energy and less thermal energy than bulbs that are less efficient. What could you measure to compare the efficiency of different light bulbs?

Efficiency of energy transformations

M1

A high-jumper in action has this energy transformation:

chemical energy → potential energy + thermal energy + sound energy

For every 100 J of chemical energy taken in (from food), the athlete changes about 20 J into useful potential energy (to gain height).

The athlete's useful energy output is 20 J. Their total energy input is 100 J.

Efficiency = useful energy output ÷ total energy input
= 20 ÷ 100 = 0.2

This means the athlete's efficiency is 20 ÷ 100 = 0.2 (or 20%)

Floodlighting should be controlled to avoid light pollution

Research

What is the impact of large sports events on the environment? You could consider pollution caused when spectators travel to and from events and pollution caused by building a large stadium.

Improving efficiency

D1

Efficient electrical or heating devices save energy, and reduce the effects on the environment. Many large sports events take place at night under floodlights. If organisers use efficient equipment, they reduce their fuel and electricity costs. This also reduces the impact on the surroundings and the environment in general. For example, well designed floodlighting means:

- more light shines directly on the courts or pitch so there is less light pollution
- more efficient lights produce less waste energy as heat
- less electricity is used, so demand for electricity from power stations is reduced
- less fuel is burned when generating electricity at the power stations.

...improving energy efficiency

Sources of energy

Can we get energy for free?

Windmills, water mills and the heat of the Sun once provided all our energy needs. Now billions of people live on Earth and more of us than ever before want to live in comfort. It's time to make decisions about where our energy is coming from.

P1

Non-renewable energy sources: fuels

Type of fuel	Description
fossil fuel	Coal, oil and gas, which are being used up more quickly than they are forming. They have many uses when they are burned – heating, cooking, generating electricity, transport.
nuclear fuel	Uranium and plutonium, which are radioactive materials found in the Earth. Their nuclear reactions give out heat, when used in nuclear power stations to generate electricity.

Renewable energy sources

Type of energy source	Description
solar energy	Energy from the Sun, which can be used directly in two main ways. **Solar cells** convert the Sun's energy into electricity. Calculators use solar cells. **Solar panels** on the roofs of buildings use the Sun's thermal radiation to heat water. This is then used in central heating and hot water systems.
wind energy	The wind spins blades on wind turbines, generating electricity.
biomass	Waste wood and other natural materials can be burned in power stations instead of fossil fuels.
biofuels	Plants are used to make fuel alcohol which can be used instead of petrol.
hydroelectricity	Large dams are built across rivers, trapping water. Gates can be opened to allow water to fall through tunnels, spinning turbines.

Our use of energy has increased dramatically over the last century

Know more

Non-renewable energy sources are fuels that took millions of years to form. Once they are used up, they cannot be replaced.

🔍 ...non-renewable energy

M1

Power stations are needed to generate large amounts of electricity at all times, whatever the weather

Which sources should we use?

All forms of energy used to generate electricity have advantages and disadvantages.

Type of energy source	Advantages	Disadvantages
fossil fuels	Generates large amounts of electricity when needed.	Causes pollution which is believed to be leading to global warming. Will run out in the future.
nuclear power	Generates large amounts of electricity when needed. No greenhouse gases emitted.	Radioactive waste needs safe disposal. High costs involved. Will run out in the future, but not as soon as fossil fuels.
solar energy	Will not run out. Useful in remote places. No greenhouse gases emitted.	No use at night or if it is cloudy.
wind energy	Will not run out. Useful in remote places. No greenhouse gases emitted.	No use if there is too much or too little wind.
biofuels	Will not run out.	Causes pollution which is believed to be leading to global warming. Biofuel crops can reduce the land available to grow food crops.
hydroelectricity	Generates electricity when needed. Large-scale schemes can provide large amounts of electricity.	Causes flooding behind the dams. Affects the normal flow of rivers.

Discuss

Sources of fossil fuels are likely to run out during your lifetime. It will then become very expensive to generate electricity on today's scale. How much are you prepared to pay for electricity? Would you prefer to change your lifestyle? Are you prepared for power cuts?

Using solar energy

Think about

Suggest how your school or college could reduce its use of energy.

Using less

D1

If we use less energy at home or at work, power stations can generate less electricity, and energy sources will last for longer. Ways to use less energy include:

- choosing energy-efficient equipment
- using equipment more efficiently.

🔍 ...renewable energy

Environmental impact

Your assessment criteria:

M1 Describe the energy transformations and the efficiency of the transformation process in these investigations

D1 Explain how energy losses due to energy transformations in the home or workplace can be minimised to reduce the impact on the environment

Saving the Earth

The cost of air travel has fallen so much that many people travel abroad several times a year. This is why a third runway at Heathrow is proposed. But this comes at a huge cost to the environment.

Is more air travel the best way to use our limited resources?

M1 Benefits of efficient energy transformations

We can reduce the energy we use by making simple changes. The benefits of using less energy are:

- less damage is done to the environment in extracting energy resources like fossil fuels

- less **pollution** is caused when burning fuels or producing energy from other sources

- less waste is produced

- resources will last for longer.

This will only happen if we make **energy-efficient choices**, such as using more efficient equipment, or using it in a more efficient way.

Using less energy will reduce the amount of coal we need to extract from mines

D1 Making energy-efficient choices

We use energy for many purposes, such as transport, electric devices and lighting, heating, cooking and manufacturing.

Transport choices

If someone needs to go to Australia for a short business trip, flying is the obvious choice. However, many companies now set up video conferences so people don't have to be in the same room to have a meeting.

People won't use public transport unless it is as cheap and as convenient as using cars. If a very small number of people use a bus service for some routes, the cost to the environment of running empty or nearly empty buses may be more than if people used their cars instead.

Research

What is meant by a 'carbon footprint'? What changes could you make to reduce your carbon footprint?

D1

Increasingly, smaller and more energy-efficient cars are being produced. Drivers must decide if it is more energy efficient to keep using an older, less efficient car or to scrap it and buy a newer more efficient car. A cost–benefit analysis for changing cars will look at:

- the cost of buying a new car

- the environmental cost of scrapping an old car and producing a new one

- the running cost of each car

- the impact on the environment of running each car.

Would scrapping the older car to buy a new one help to save the environment?

Choices in electricity production

Generating electricity from fossil fuels and nuclear power means we have a reliable supply of electricity at all times. However, fuels need to be extracted and transported to the power stations before use, and there are limited supplies. The pollution caused by burning fossil fuels, and problems with dealing with waste products from fossil fuel and nuclear power stations, means that renewable energy sources are increasingly being used.

Renewable energy sources cannot produce energy all the time or in all places, and are not always cost effective. Many small units such as wind turbines and solar cells are needed, pushing up the cost of the electricity generated. A back-up supply of energy is also needed for the times when renewable sources cannot produce energy.

A **cost–benefit analysis** for a new power station will look at:

- the economic and environmental costs of building each type of power station

- the economic and environmental costs of running each type of power station

- the economic costs and environmental risks of waste disposal for each type of power station

- the reliability of the method of electricity generation.

The environmental impact of this offshore wind farm is low but its cost is high

Research

The cost per unit of electricity produced is a useful comparison for different types of power stations. Try to find out some comparative data and comment on what you find out.

Waves and communication

Your assessment criteria:

P3 Describe the electromagnetic spectrum

P5 Describe how waves can be used for communication

M3 Explain the advantages of wireless communication

D3 Compare wired and wireless communications systems

Passing the message on

Hundreds of years ago, ships at sea used flags and flares to send long-distance messages. Nowadays, radio signals are used to communicate with other ships, with the shore, with satellites and with submerged submarines. Information is carried almost instantly using radio waves travelling between the transmitter on the ship and the receiver.

P3 — What are waves?

Waves carry energy from place to place. Imagine waves travelling through a tank of water. They carry energy as they travel across the tank, and can make an object at the far end of the tank bob up and down. Many waves behave in a similar way to water waves. The energy is carried through different materials, even though the material itself doesn't travel. There are different types of wave.

Sound waves are caused by vibrations, and travel though solids, liquids and gases.

Electromagnetic waves are a group of waves that includes radio waves and light waves. All electromagnetic waves:

- travel through empty space (a vacuum) as well as through solids, liquids and gases

- travel at the same speed in a vacuum (300 000 km/second).

All waves have the following features:

- The **wavelength** is the distance between peaks. It's also the distance between troughs. Wavelength is measured in metres.

- The number of waves passing the same place every second is **frequency**. Frequency is measured in **hertz** (Hz).

- The **wave speed** measures how quickly a wave carries energy from one place to another. Wave speed is measured in metres per second (m/s).

wave speed = wavelength × frequency

Know more

Sound waves travel about a million times more slowly than light waves. If a car door slams shut some distance away, you hear the noise a little time after you see the door close. That's why you hear thunder after you see lightning.

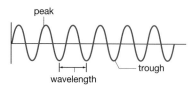

High frequency wave

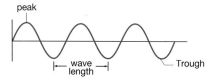

Low frequency wave

 ...wave speed

Know more

Here is an example of calculating wave speed:

The frequency of a wave is 10 Hz. Its wavelength is 5 m.

Speed = frequency
× wavelength

$$= 10 \times 5 = 50 \text{ Hz}$$

Radio waves help keep emergency and security services in touch

Communications

P5

The first international telephone systems used copper wires which connected to places in different countries, along the sea bed where necessary. Now, many long-distance telephone connections and internet links are wireless. Radio waves and microwaves send signals to satellites, which transmit them onwards.

To send a signal, you need a **transmitter**. To receive a signal you need **receiver**.

Type of communication	Example of transmitter	Example of receiver
changing TV channels	remote controller	TV set
mobile phone call	mobile phone mast	mobile phone
satellite TV programme	transmitter on satellite	satellite dish on house

Advantages of wireless communication

M3

Wireless communication, using radio and microwaves, has advantages:

- the system isn't fixed in one place
- there are no wires that can be damaged
- it's easy to change or add more users.

There are also however some disadvantages:

- security is a problem as it is easier to hack into the system
- the signal can deteriorate.

Wireless or wired?

D3

People can choose:

- satellite TV (wireless) or cable TV (wired)
- mobile phone (wireless) or a landline (wired)
- internet connections that are wireless or wired.

The choice depends on what is important for the user. Security may be an issue, or the most important thing may be the ability to move round, or the cost.

The electromagnetic spectrum

Your assessment criteria:

P3 Describe the electromagnetic spectrum

P4 Describe the different types of radiation including non-ionising and ionising radiation

M2 Describe the uses of ionising and non-ionising radiation in the home or workplace

D2 Discuss the possible negative effects of ionising and non-ionising radiation

Waves from the Sun

The Sun gives out a range of electromagnetic waves which affect us in different ways. Visible light allows us to see, and allows plants to grow; our skin senses infrared waves as heat; our skin also absorbs ultraviolet radiation, and gradually tans. X-rays and gamma radiation from the Sun are mainly absorbed by the Earth's atmosphere, but can be detected in aeroplanes flying high.

P3 The electromagnetic spectrum

Electromagnetic waves are a family of waves that carry different amounts of energy depending on their frequency. Waves with a low frequency carry less energy than waves with high frequency, and so they have different uses.

All electromagnetic waves travel at the same speed (in a vacuum). Since speed = wavelength × frequency,

- high-frequency waves have a short wavelength
- low-frequency waves have a long wavelength.

Visible light is one type of electromagnetic wave

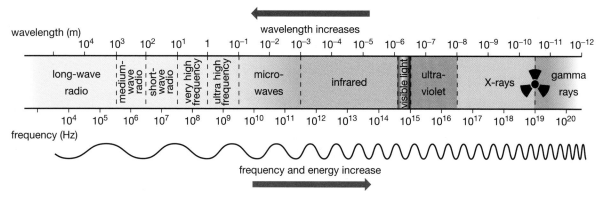

The electromagnetic spectrum

P4

We put electromagnetic waves into groups depending on their frequency and wavelength. Together they make up the **electromagnetic spectrum**. Low frequency waves like radio waves, microwaves and infrared are non-ionising. High frequency waves like x-rays and gamma radiation are ionising.

...electromagnetic spectrum

Think about

Which electromagnetic waves have you used today?

Using electromagnetic waves

We can make use of electromagnetic waves in many different ways, depending on their properties. The properties vary, depending on the frequency and wavelength.

Type of wave	Properties	Uses
radio waves (non-ionising)	Can be reflected off an electrically charged layer in the Earth's atmosphere.	Used to broadcast TV and radio programmes.
microwaves (non-ionising)	Pass through the atmosphere easily. Absorbed by water, heating it up.	Used to communicate with satellites and mobile phone networks. Used to cook food.
infrared waves (non-ionising)	All objects give out some infrared waves. Can be sent as a narrow beam.	Used in grills, toasters and electric fires. Used in TV remote controls.
visible light	Can be sent as a narrow beam. Detected by our eyes.	Our eyes see visible light reflected off objects, or given out by glowing objects.
ultraviolet radiation	Absorbed and re-emitted as visible purple light from fluorescent materials. Tans our skin, but damaging in excess.	Used to detect bank note and ticket fraud. Used in sunbeds.
x-rays (ionising)	Absorbed by bones and metal but pass through skin and cloth. Damaging to cells in excess.	Used in x-ray imaging to detect broken bones and cracks in metals by creating shadow pictures.
gamma rays (ionising)	Kill cells.	Used to treat cancer by killing cancer cells. Used to sterilise equipment by killing bugs.

Increasing frequency →

Increasing wavelength →

Over-exposure to ultraviolet not only leads to sunburn but also increases the risk of skin cancer

Research

Find out about the effect of the thinning of the ozone layer in the atmosphere. How have ozone holes affected numbers of people developing skin cancer?

When are electromagnetic waves harmful?

Some electromagnetic waves can cause damage to human cells, depending on the type of wave (that is, its frequency), how long you are exposed to it, and how intense (strong) it is.

High frequency electromagnetic waves are ionising and can damage or kill cells. The use of gamma radiation and x-rays is carefully controlled as high doses can cause radiation sickness and cancer. Ultraviolet radiation is ionising. Ultraviolet rays from the Sun can cause sunburn and skin cancer.

Low frequency electromagnetic waves are non-ionising but can still cause harm. When radio waves, microwaves and infrared waves are absorbed, their energy causes cells to heat up. If the radiation is intense, it can cook cells, killing or damaging them. Low levels of non-ionising radiation such as microwaves from mobile phones possibly may cause brain tumours.

Anyone exposed to ionising or non-ionising radiation should limit the time or intensity of their exposure to reduce harm.

...ozone layer and skin cancer

Ionising radiation

Your assessment criteria:

P4 Describe the different types of radiation including non-ionising and ionising radiation

M2 Describe the uses of ionising and non-ionising radiation in the home or workplace

D2 Discuss the possible negative effects of ionising and non-ionising radiation

Radiation and cancer

Cancers are caused when damaged cells in the body replicate rapidly, forming a tumour. Smoking and exposure to ionising radiation are two of the factors that increase the risk of developing cancer.

But often, people with cancer have radiotherapy – ionising radiation is used to cure the cancer it may have caused. How is that possible?

P4 Types of ionising radiation

Everything is made up from tiny particles called atoms. Atoms have a central **nucleus** surrounded by tiny **electrons**.

Radioactive elements have an unstable nucleus. They give out **ionising radiation** when their nuclei break up. Ionising radiation affects other atoms by knocking electrons from them.

The following types of ionising radiation are given out by radioactive elements:

Alpha particles are helium nuclei – helium atoms without the electrons.

- They are very ionising.
- They travel only a short distance in air.
- They cannot pass through skin or paper.
- They are very damaging inside the body.

Beta particles are electrons.

- They are less ionising than alpha radiation.
- They have high penetrating power and pass through skin.
- 5 mm of aluminium sheet will stop them.

Gamma rays are high-energy (high-frequency) electromagnetic waves.

- They are not as ionising as beta or alpha radiation.
- They can penetrate almost everything.
- A block of lead about 10 cm thick will only stop half of them.

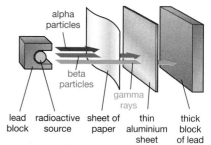

The three radiations and their penetrating abilities

 ...ionising gamma rays

Using ionising radiation

Use	Type of radiation	How it works
smoke alarm	alpha	Normally, the air in the alarm is ionised by alpha particles, causing a current to flow. Smoke absorbs alpha particles, stopping the current and setting off the alarm.
paper thickness monitoring (during manufacture)	beta	Beta radiation passes through the paper being produced on rollers. If the paper is too thick, less radiation is detected so the rollers move closer. If the paper is too thin, more radiation is detected and the rollers move apart.
treating cancer	gamma	Fine beams of gamma radiation are aimed at the tumour, killing the cancer cells.
tracing leaks	gamma	To look for a leak, a sample of radioactive material is put into the liquid. A radiation detector shows where the fluid collects. This works for blood in patients, oil in engines and liquids in underground pipes.
sterilisation	gamma	Harmful bacteria on surgical equipment are killed by gamma radiation. This works with food too – the radiation kills bacteria that would make the food go off.
gamma scan	gamma	A radioactive liquid is injected into a patient's blood. A gamma camera detects where more of the radioactive blood collects, which may indicate a tumour.

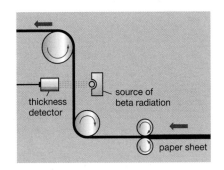

Controlling paper thickness

Discuss

Should we worry about radioactivity and cancer? Find out the main causes of cancer. Is radioactivity the main risk factor? Which causes can you control with lifestyle choices?

Problems with ionising radiation

Sometimes ionising radiation damages living cells. The damage depends on:

- how strong the source is
- where the source is in relation to the person
- how long a person is exposed to it
- the type of radiation.

Protective clothing limits the dose of radiation we receive

Low doses, or higher doses for very short times, are unlikely to cause problems.

Alpha radiation is only harmful inside the body (if eaten, injected or breathed in).

Gamma radiation is highly penetrating, so sources outside the body can be very harmful.

...radiation uses

Circuits

Your assessment criteria:

P8 Describe the use of measuring instruments to check values predicted by Ohm's law in given electrical circuits

D4 Assess how to minimise energy losses when transmitting electricity and when converting it into other forms for consumer applications

Circuits all around us

All around us electrical equipment entertains and informs us, communicates, makes decisions, carries out work, and does a whole range of different jobs. Battery-operated equipment has a complete circuit inside it. Plugs are used to join mains-operated equipment to a circuit linking the home to the power station where electricity is generated.

P8

What is a circuit?

All electrical equipment contains circuits. To work, every circuit needs:

*   a **power source** (battery or mains electricity)

*   a continuous loop of **electrical conductors** (usually plastic-covered copper wire)

*   **components** (such as a bulb) that change electrical energy into different forms.

Circuits are drawn using symbols to represent the different parts.

When a circuit is switched on, there must be no gaps in the loop of conductors. Then:

*   tiny charged particles (electrons) flow round the circuit from the power source

*   the electrons carry energy from the power source to different components in the circuit.

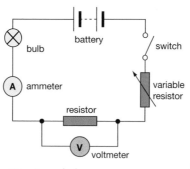

Circuit symbols

The flow of electrons is called a **current**. Current – rate of flow of charge – is measured in **amps** (A) using an **ammeter**. The ammeter is connected in the circuit so that the current through it is measured.

Know more

Electrical conductors are materials that let electricity flow through easily, for example metals and graphite. Electrical insulators, like plastic and wood, have a high resistance to current.

Measuring current and voltage in a circuit

...circuit symbol charts

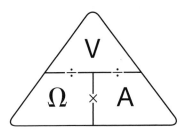

You can use this triangle if you are asked for the resistance or current. Cover up the quantity you need to find and the rest of the triangle shows you how to calculate it

P8

The **voltage** (also called **potential difference**) is a measure of the energy carried by the current. Voltage is measured in **volts** (V) using a **voltmeter**. The voltmeter is connected across the circuit – in parallel with the component(s) – because it is measuring the difference in energy between two points in a circuit.

Resistance

Resistance measures how hard it is for a current to pass through a material. Resistance is measured in **ohms** (Ω). It is calculated by measuring the current that flows when different voltages are applied in the circuit.

For many materials, doubling the voltage doubles the current. These materials obey Ohm's law, which states:

voltage (volts, V) = resistance (ohms, Ω) × current (amps, A)

Here is an example of using Ohm's law.

The resistance of a bulb is 30 ohms. If a current of 0.2 amps flows through it, what is the voltage across it?

Resistance = 30 Ω   Current = 0.2 A

Voltage = resistance × current

$$= 30 \times 0.2$$

$$= 6 \text{ V}$$

D4

Minimising energy losses

Current passes easily through some wires but not others. More energy is needed to push electrons through:

- longer wires

- thinner wires

- wires made from materials that are not such good conductors.

Copper is used in circuits throughout the home because it has a low resistance. This means that, compared with wires that are not such good conductors:

- less energy is needed to push electrons through the wires

- more energy is available for equipment that is plugged in.

Think about

Why is it important to use highly conducting copper for electrical wiring? What would be the effects of using material of a higher resistance? Why are copper wires coated with plastic?

...electrical circuits

Where our electricity comes from

Over 150 years ago, a scientist called Michael Faraday discovered that moving a magnet near a coil of wire created an electrical voltage in the wire. Today, this simple idea is used in every power station, generator and dynamo to generate the electricity we take for granted.

P6

Generating electricity

Magnetism, electric current and movement of a conductor always go together. Wherever you find two of them, you will find the third as well.

Push a magnet into a coil of wire and a voltage is created in the wire. The same thing happens if the magnet is pulled out of the coil or if it spins inside it.

If the coil is part of a circuit, then the voltage forces a current to flow in the wire. The current can be measured using an ammeter, or used to light a bulb.

A larger voltage is created if:

- the magnet moves more quickly

- a stronger magnet is used

- there are more turns on the wire coil.

It doesn't have to be the magnet that moves – a voltage is also created if the coil moves towards or away from the magnet.

The generator

A generator is designed to create electricity by spinning the coil of wire or the magnet. In the simple generator shown here:

- the magnets are fixed

- a handle or wheel spins the coil of wire

- slip rings are used so that the wires don't get tangled as the coil spins

- carbon brushes are conductors that connect the rest of circuit to the generator.

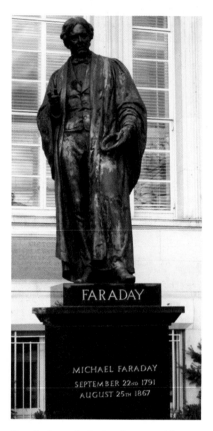

Faraday's work changed our world

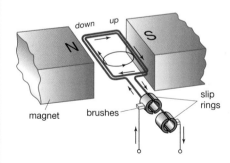

This generator can easily be made in a school lab, but the same ideas are used in power stations

...how a generator works

Know more

Even before electricity was discovered, rotation was used to capture energy and make it useful, in windmills and water mills.

Know more

Recently built gas-powered generating stations have two stages:

- *gas is burned, producing jets of hot gas which spin a turbine*

- *the hot gases also heat water, producing steam which turns another turbine.*

This improves the efficiency of electricity generation.

Think about

What would you look for if you were choosing an energy source? Rank these in order of importance: cost of fuel; cost of the generating equipment and buildings; reliability; efficiency; pollution caused; how easily available or transportable the energy source is.

How to make the generator spin

All power stations use a **turbine** to spin the generator. A turbine has blades (like a windmill). The turbine spins when jets of steam, moving water or the wind pass over its blades. Turning the turbine turns **electromagnets** in the generator.

The difference between different types of power station is what makes the turbine spin.

In power stations fuelled by non-renewable energy sources, fossil fuels or nuclear reactions create heat. This changes water to steam which passes through the turbine making it spin.

In generators powered by renewable energy sources:

- wind turns the blades of a wind turbine directly

- moving water in hydroelectric schemes and tidal schemes turns turbines directly

- biomass or biofuel heats water in a boiler when burned, and the steam produced spins the turbine.

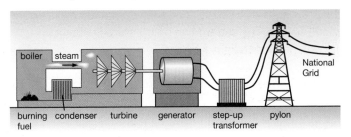

Different stages in a power station

Comparing the efficiency of power stations

M4

Remember that efficiency is the proportion of energy input that is transformed into useful energy, in this case electrical energy. Different methods of electricity generation have different levels of efficiency.

Type of power generation	Efficiency
fossil fuel (coal, oil or gas)	35%
recent 'combined cycle' gas turbines	60%
nuclear	35%
hydroelectric	80%
wind	20%
biofuel	20%
biomass	35%

...how to generate electricity

Electrical energy

Your assessment criteria:

P6 Describe how energy can be produced

P7 Describe how electrical energy is transferred to the home or industry

D4 Assess how to minimise energy losses when transmitting electricity and when converting it into other forms for consumer applications

Getting the electrical energy to where we need it

Over 700 million batteries are bought in the UK each year. Batteries are useful because they provide a portable source of electricity. The choice ranges from tiny batteries for hearing aids to very large batteries for cars, with 'AA' batteries one of the most popular types. It is likely that in every room in your house there is at least one battery-operated item.

Know more

*What you are used to calling a 'battery' is actually an electric **cell**. A battery is, strictly, a collection of cells in a circuit.*

P6 Using batteries

Batteries transform chemical energy to electrical energy through chemical reactions. They provide a steady current called **direct current**. They can be **non-rechargeable** or **rechargeable**.

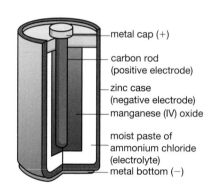

metal cap (+)

carbon rod (positive electrode)

zinc case (negative electrode)

manganese (IV) oxide

moist paste of ammonium chloride (electrolyte)

metal bottom (−)

Structure of a battery

- In a non-rechargeable battery, once the chemical reaction is completed the battery needs to be disposed of, and cannot be reused. Most batteries we use are non-rechargeable.

- Rechargeable batteries can be used many times before they need to be disposed of. When they run down, they are plugged into a mains supply of electricity using a charger. This reverses the chemical reaction inside the battery using electrical energy from the mains.

- You can buy chargers for rechargeable batteries, or you can plug equipment such as mobile phones and cameras into the mains for recharging their batteries, or you can place some devices such as a cordless drill into a special base unit to recharge.

Know more

Batteries contain heavy metals which are poisonous if they leak into the ground or rivers. Almost all batteries can be recycled safely and yet most people don't bother.
From February 2010, more retailers need to provide places for people to recycle their used batteries, so the recycling rate should improve.

🔍 ...recycling batteries

Using mains electricity

P7

Electricity is generated in power stations. Then it has to get from the power stations to our homes along a network of cables, pylons and transformers called the **National Grid**.

This electricity is **alternating current**. It isn't a one-way flow, like current from a battery, but it changes direction repeatedly, 50 times a second. Alternating current is easier to generate and can be transmitted more efficiently over long distances.

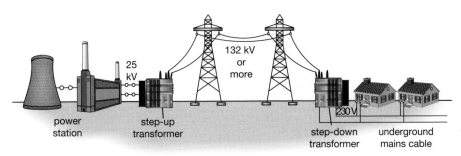

power station | step-up transformer | 132 kV or more | step-down transformer | underground mains cable | 25 kV | 230 V

Electrical energy from power stations reaches our homes via the National Grid

Think about

Why is it of concern that most wind farms are in remote places?

Transmitting electricity efficiently

D4

Transmitting electricity over long distances causes energy losses as the cables heat up.

Transformers are used to increase, or **step up**, the voltage to many thousands of volts before the electricity is output to the National Grid. This reduces the current, so the wires do not heat up as much. This means less energy is lost as thermal energy.

Near our homes, transformers in **sub-stations** reduce, or **step down**, the voltage to 230 volts so it is at a suitable level for our appliances and a safer level for us to use.

Another way to reduce energy losses when transmitting electricity is to reduce the distance over which it is transmitted.

Research

Try to find out what percentage of energy is lost in the National Grid.

Transformers at a sub-station change the voltage in the cables

Our solar system

Shooting stars

At certain times of the year, showers of shooting stars can be seen in the night sky. Fast-moving intense streaks of light are caused by pieces of rock burning as they race through the night sky, so bright they look like stars. These 'shooting stars' are meteors. Sometimes they hit the ground as rocks called meteorites.

P9 The solar system

Our solar system contains these parts.

- The Sun is the **star** at the centre of the solar system. The heat and light from the Sun is caused by nuclear reactions in its centre.

- **Planets** are large balls of rock or gas that travel around the Sun in nearly circular orbits. There are eight planets in our solar system, and Earth is one of them.

- **Dwarf planets** also orbit the Sun, but are smaller than planets. They have less effect on their surroundings than planets. Pluto is a dwarf planet.

- **Moons** are smaller lumps of rock that travel around planets. Some planets have several moons.

- **Comets** travel through space and are formed from small lumps of rock and ice. Some orbit the Sun in long, stretched-out orbits, so that the comet is sometimes far from the Sun and at other times much closer to the Sun. When it is close, the comet appears to have a tail streaking out from behind it.

- **Asteroids** behave like planets and dwarf planets, but are much smaller. Most are found orbiting between Mars and Jupiter.

- **Meteors** are also called shooting stars. They occur when rocks burn up as they pass through the Earth's atmosphere.

Know more

The four inner planets have a rocky surface you could walk on. The four outer planets, called the gas giants, are very massive and formed from gases – you would sink into them if you tried to walk on them. The rocky planets are much closer to one another than the gas giants.

Parts of our solar system

...solar system

How was our solar system formed?

M5

Stars form from swirling clouds of dust and gases in space called **nebulae**.

- Gravity makes the centre of the cloud clump together.
- This clump is so massive that gravity crushes the particles very, very tightly.
- The centre heats up to millions of degrees.
- This is hot enough for nuclear fusion reactions to ignite.
- The star starts to shine and gives out heat.

As the star forms, the outer edges of the cloud are still swirling around the centre. These particles of dust and gas may clump together, attracting other nearby particles. Their gravity is not strong enough for nuclear reactions to ignite, but it is strong enough to hold the particles together. This is how planets are formed. Their movement means that they orbit the star, and the solar system is complete.

The **universe** is the whole of space and includes billions of groups of stars called **galaxies**, each with billions of stars, many with a solar system. Our own galaxy is called the Milky Way.

During a star's lifetime, it changes size and colour as different reactions take place in its core. Eventually, some stars explode, flinging out elements that have formed in the nuclear reactions in their core. These particles form nebulae, where new stars will form.

A nebula where new stars are forming

Research

Have you heard of a red giant, and a white dwarf? Research the main stages in the life cycle of a star and find out what these names mean.

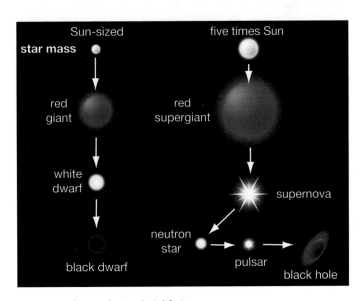

How stars change during their lifetime

...star life cycle

Your assessment criteria:

P10 Identify evidence that shows how the universe is changing

M5 Describe the main theory of how the universe was formed

M6 Explain how the evidence shows that the universe is changing

D5 Evaluate the main theory of how the universe was formed

D6 Evaluate the evidence that shows how the universe is changing

Investigating starlight

When different chemicals burn, the flames change colour. An instrument called a spectroscope is used to analyse the light coming from different elements when they are heated. Each element has its own special **spectrum** made up of lines of different colours. If scientists look at light coming from far-away stars, they can tell what elements are in the star, giving out light in its hot core.

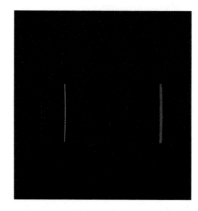

The spectral pattern for hydrogen

P10 What shows us that the universe is changing?

Red shift

Scientists studying light from far-away galaxies have found that all the lines in the spectra were redder than expected. This is called the **red shift**. The red shift is bigger when a galaxy is further from us. Analysis of red shift told scientists that all galaxies were moving away from us, and the most distant galaxies were moving away fastest. In other words, the universe is expanding.

Cosmic background

About 40 years ago, scientists detected radiation coming, very faintly, from all directions of space. They called it **cosmic background radiation**. This discovery proved very important, as it confirmed the theory that the universe had begun as a 'big bang'.

Observations

Telescopes like the Hubble Space Telescope take photographs of stellar explosions and nebulae, providing evidence that throughout the universe stars are dying, and new stars and galaxies are constantly forming.

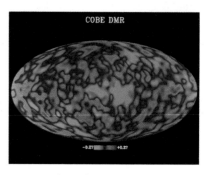

Scientists have detected cosmic background radiation

Know more

The scientists who first detected cosmic background radiation thought the signal was interference in their equipment. They even cleared out pigeon droppings from their detectors, thinking this might have been the cause.

 ...big bang

M5
M6

Discuss

What will happen to the universe in the future? Discuss these three possibilities:

- *the universe will keep on expanding*
- *the expansion will gradually slow down and the universe will reach a fixed size*
- *the universe will stop expanding and then start to contract, ending up in a 'big crunch'.*

What does the evidence tell us about the origin of the universe?

Scientists believe that the red shift evidence tells us that at some point in time, billions of years ago, all the matter and energy in the universe was concentrated in the same place. A massive explosion took place, called the Big Bang. It flung all the matter and energy apart. As the new universe cooled, particles formed and clumped together, making stars and galaxies. Ever since then, the universe has been expanding and cooling down.

If there had been such an explosion billions of years ago, scientists knew they should be able to detect an 'echo' of it today. This would be very faint radiation coming from all directions. That is why the discovery of the cosmic background radiation was so exciting and important.

a fraction of a second after   three minutes after   300 000 years after   500 million years after

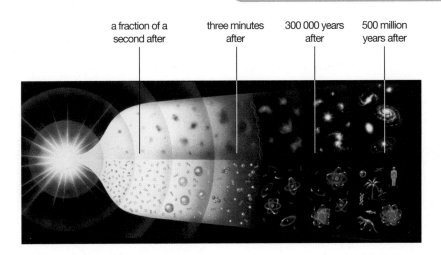

The Big Bang theory suggests that before the Universe existed, everything was concentrated in a space smaller than a pin head

We can literally see the evolution of the universe with the Hubble Space Telescope

D5
D6

Is the theory right?

The startling discoveries of the red shift of distant galaxies and the cosmic background radiation suggest that the Big Bang theory is right. There is no strong counter-evidence to say it is wrong. It cannot be proved, but most scientists believe the Big Bang took place. What we probably will never know is what happened before the Big Bang or what caused it.

The evidence that the universe is still changing is strong. Photographs and other detection equipment can record the changes clearly. They can not only tell us what is happening now, but what happened in the past. Light from stars takes years to reach us. Light from stars in the next galaxy takes thousands of years to reach us, and we have even detected light from stars formed soon after the Big Bang that has taken 14 billion years to reach us.

🔍 ...red shift theory

To achieve a pass grade, my portfolio of evidence must show that I can:

Assessment Criteria	Description	✓
P1	Carry out practical investigations that demonstrate how various types of energy can be transformed	
P2	Calculate the efficiency of energy transformations	
P3	Describe the electromagnetic spectrum	
P4	Describe the different types of radiation, including non-ionising and ionising radiation	
P5	Describe how waves can be used for communication	
P6	Describe how electricity can be produced	
P7	Describe how electrical energy is transferred to the home or industry	
P8	Describe the use of measuring instruments to check values predicted by Ohm's law in given electrical circuits	
P9	Describe the composition of the solar system	
P10	Identify evidence that shows how the universe is changing.	

Wind and solar energy

To achieve a merit grade, my portfolio of evidence must show that I can:

Assessment Criteria	Description	✓
M1	Describe the energy transformations and the efficiency of the transformation process in these investigations	
M2	Describe the uses of ionising and non-ionising radiation are used in the home or workplace	
M3	Explain the advantages of wireless communication	
M4	Compare the efficiency of electricity generated from different sources	
M5	Describe the main theory of how the universe was formed	
M6	Explain how the evidence shows that the universe is changing.	

To achieve a distinction grade, my portfolio of evidence must show that I can:

Assessment Criteria	Description	✓
D1	Explain how energy losses due to energy transformations in the home or workplace can be minimised to reduce the impact on the environment	
D2	Discuss the possible negative effects of ionising and non-ionising radiation	
D3	Compare wired and wireless communication systems	
D4	Assess how to minimise energy losses when transmitting electricity and when converting it into other forms for consumer applications	
D5	Evaluate the main theory of how the universe was formed	
D6	Evaluate the evidence that shows how the universe is changing.	

LO

Be able to investigate the impact of human activity on the environment

- With the major growth of industry in the last fifty years, some scientists believe that we may have reached a 'tipping point', where even if we reduce pollution, it will be too late to stop climate change

- Farmers in the UK are introducing methods to reduce the loss of habitats and species of animals and plants

- The 10 million tonnes of rubbish we recycle every year reduces pollution of the environment by these wastes and helps to preserve natural resources

LO

Know the factors which can affect and control human health

- Our immunisation programme helps to reduce diseases caused by micro-organisms

- Cancer is caused by chemicals, radiation, micro-organisms and our lifestyle

- Some diseases are inherited; some are caused when the body's immune system attacks itself; psoriasis is an example of an autoimmune disease

- The nervous system and endocrine system coordinate the actions of the human body

LO

Be able to investigate the functioning and classification of organisms

- We may be able to use stem cells to replace damaged or defective cells in the human body and in future use this method to treat and prevent diseases that are currently incurable

- We have around 30 000 genes on our 23 pairs of chromosomes that give us our unique identity

- All of the two million different types of known living organisms, no matter how large or how small, interact in some way with others

- It's estimated that about one-third of plants that provide us with food are dependent on bees for pollination

Cells and tissues

Cells and life

Scientists estimate that the human body contains around 100 trillion (10^{14}) cells. Our cells, and those of other organisms that are important to us, are studied by many types of scientist. They may examine cells to check that an organism is growing, developing and functioning normally.

P1

The structure of cells

All plants and animals, and most living organisms, are made up of cells. The structures of a simple animal cell and a simple plant cell are shown here.

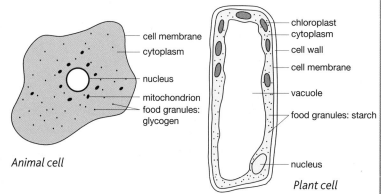

Animal cell

Plant cell

The **cell membrane** controls what substances enter and leave the cell. Chemicals such as water and food molecules enter the cell, and the cell's waste products leave.

The plant cell wall is made from a chemical called cellulose. It gives a plant cell its rigid shape.

The cytoplasm is the jelly-like material that fills the cell. It's where the chemical reactions that keep the cell alive take place.

Mitochondria are tiny structures in the cytoplasm where respiration takes place.

The plant cell vacuole is a large, permanent, fluid-filled space. The vacuole helps to keep plant cells firm.

The nucleus is the control centre of the cell. It contains tiny, thread-like structures called **chromosomes**. These carry genetic information.

Know more

Cells vary in size. Egg cells are one of the largest types of cell. And of these, an ostrich egg is the largest, measuring on average around 150 by 130 mm.

Nerve cells are longer than this, but much, much narrower. The nerve cells of the giant squid can be 12 m long.

Know more

The number of chromosomes in the nucleus of cells varies from one organism to another. The male of one species of ant has only one chromosome, while one type of fern has over a thousand. Humans have 23 pairs of chromosomes.

The female jumper ant has just one pair of chromosomes; the male ant only one chromosome

 ...plant and animal cell structure

Know more

Scientists think we may be able to use stem cells to repair the human body and treat diseases. In 2009, stem cell treatment was used to restore the sight of someone who was partially blind. But culture of stem cells from human embryos is controversial.

Different types of cells

P1

As a foetus develops, its cells become specialised to do certain jobs. The human body contains about 200 different types of cells. Some of the most important types are shown here.

Stem cells remain in an unspecialised state. These have the potential to develop into any type of cell in the body.

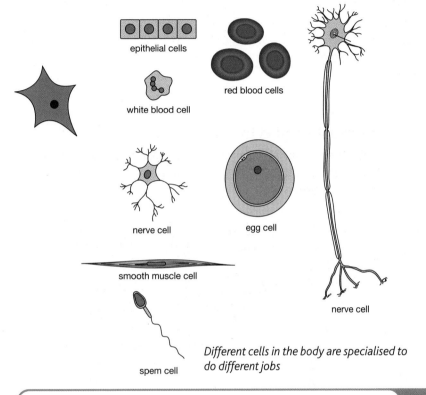

Stem cells

Different cells in the body are specialised to do different jobs

Tissues

P1

Most cells in the body don't work in isolation. They work closely together. Different types of cell are organised into **tissues**.

Tissue	Role in the body
blood	transports substances and helps to fight disease
connective	supports organs and binds tissues together
epithelium	covers or lines organs of the body
muscular	contracts to move parts of the body
nervous	transmits messages around the body
reproductive	produces sex cells
skeletal	supports the body and makes it rigid; includes bone and cartilage

Know more

In the school lab, you will look at simple plant and animal cells and tissues with a light microscope. Scientists also use light microscopes to examine cells and tissues, but if they use an electron microscope, they can work at higher magnifications and see more detail.

...types of human tissues

The genetic code

Your assessment criteria:

P1 Describe how the functioning of organisms relates to the genes in their cells

D1 Explain how genes control variation within a species using a simple coded message

The human genome project

The human **genome** project was a study to map the location of all the genetic information on the chromosomes of a human being. Work started in 1990. The final map was revealed in April 2003. By knowing where genetic information is located, scientists can learn more about how our cells function. The information will also give us a greater understanding of human disease.

P1 — The structure of chromosomes

Chromosomes are made of a chemical called deoxyribonucleic acid, or DNA. The shape of the DNA molecule is called a double helix.

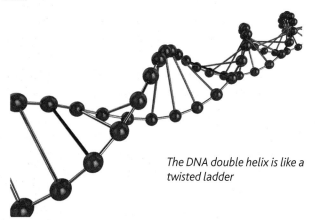

The DNA double helix is like a twisted ladder

The DNA carries a code – the genetic code. This code makes you the type of organism you are – a human – and *who* you are.

A set – or karyotype – of human chromosomes

Know more

The rungs of the DNA ladder are made of chemicals called bases. The four types of base – adenine, thymine, guanine and cytosine – are given the letters A, T, G and C.

D1 — How do genes work?

A section of DNA that carries the code for a particular characteristic is called a **gene**. Genes control the activities inside our cells, and so how cells function. They do this by controlling the production of **proteins**. Some of the most important proteins are enzymes. These control the rate at which chemical reactions work inside (and outside) our cells.

Our 30 000 genes carry the codes to control *all* our inherited characteristics. These characteristics range from things like our height, hair and eye colour, down to the minutest detail in our cells.

...DNA structure

D1

You inherit some of your characteristics from your father, and some from your mother. You will pass some of these on to your children. Unless you have an identical twin, your genes are unique to you. This uniqueness of our genes explains why we are all different. It's the cause of **genetic variation**.

The structure of proteins

Our bodies need many different types of protein. For example, they may be:

- enzymes
- structural proteins that make up our hair, muscle and bone
- hormones
- antibodies in our blood.

These proteins are very different in structure. But they are all made up of amino acids. There are 21 different amino acids. It's the number and combination of these amino acids, and the order in which they're arranged in the protein, that's important.

How do genes control the production of proteins?

When a protein is made, genes provide the code for how it is assembled from amino acids.

The four bases in DNA – A, T, C and G – work in threes, called base triplets. Each base triplet is the code for a single amino acid. The amino acids are assembled to make a particular protein in a process called **translation**.

The amino acids in hair protein form a helix

The amino acid chain in haemoglobin, the protein in red blood cells, is very folded

The sequence of bases defines the order in which amino acids are assembled into a protein

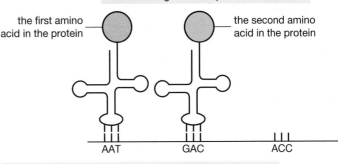

2 Amino acids are ferried in and assembled according to the sequence of bases

the first amino acid in the protein

the second amino acid in the protein

AAT   GAC   ACC

1 A single-strand template of the bases in the gene is made

Discuss

In future, it may be possible for all of us to have a copy of a map of our genome. The information could tell us if we were likely to develop a particular disease, or to die early. If insurance companies had this information, they might not want to sell us life insurance. Would you want to know this information from your genome?

The variety of life

Discovering new organisms

New species of plants and animals are being discovered every day. In 2009, scientists discovered 40 new species of animal in just one expedition to a rainforest in the crater of an extinct volcano in Papua New Guinea.

This transparent species of frog, called a glass frog, is a new animal species reported in January 2010. Its heart can be seen through its transparent chest

Know more

So far, scientists have identified and named almost 2 million species of organisms. But they estimate that there may be anything from 10 million to a 100 million different species on the planet. Over half of these are likely to be different types of insect.

P2 Classifying organisms

When a new species is discovered, scientists **classify** it. They look for distinctive features of the organism. They aim to place it in a group along with other organisms having similar features.

All living organisms are grouped into five large groups called **kingdoms**. The kingdoms are sub-divided into smaller groups, and these groups into smaller groups still.

| Kingdom |
| Phylum |
| Class |
| Order |
| Family |
| Genus |
| Species |

Kingdoms are divided into smaller groups

Some fungi are single-celled, like yeast. Others, such as moulds, mushrooms and toadstools, are made up of tiny threads of cells called hyphae

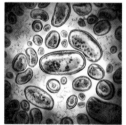

Bacteria are very small, simple cells. They have a cell wall. Their DNA is not enclosed in a nucleus

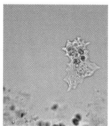

Protists are single-celled organisms, such as Amoeba and certain algae

Plants are multicellular. The cells above ground contain chlorophyll. Non-flowering plants reproduce using spores or cones. Flowering plants produce flowers

Groups of living organisms

Animals are multicellular. Most feed using a mouth. Invertebrates are animals without backbones. Vertebrates are animals with backbones: fishes, amphibians, reptiles, birds and mammals

...classification of organisms

Know more

Viruses are different from all other living things. They are not made up of cells. They have DNA (or a related chemical, RNA) enclosed in a protein coat.

Know more

The classification system we use is based on the one devised by Carl von Linné in the 1700s. It gives all organisms a name in two parts. Humans have the scientific name Homo sapiens; 'Homo' is the **genus** *(shared with some of our fossil ancestors), 'sapiens' the* **species** *(specific to us).*

Identifying organisms

When studying living organisms in their habitat, it's important to identify them. You *could* work through a pocket guide with photographs or drawings until you make a correct identification. But most biologists would use an **identification key**, especially when the organism is difficult to identify.

An identification key is a simple pathway to identifying an organism. There are questions that give you two choices. You must decide which of these choices matches your organism. Then you go to the next question. You continue until you identify the organism. There is an example below.

1	Is the animal's body divided into parts called segments?	go to question 2
	Is the animal's body not divided into parts called segments?	go to question 3
2	Do the animal's legs have joints?	crab
	Do the animal's legs have no joints?	marine worm
3	Has the animal a shell?	go to question 4
	Has the animal no shell?	go to question 5
4	Is the shell in two parts?	cockle
	Is the shell in one piece?	top shell
5	Has the animal tentacles?	sea anemone
	Has the animal five arms?	starfish

You can identify these organisms using the key above

🔍 ...identification keys for organisms

Variation and evolution

Your assessment criteria:

P3 Describe the interdependence and adaptation of organisms

M1 Describe how variation within a species brings about evolutionary change

Adaptation to the environment

Polar bears are adapted to living on Arctic ice. Their thick white fur provides insulation and camouflage, and a layer of fat provides further insulation. Scientists are currently studying animals such as polar bears very closely. They are concerned about what might happen as their environment changes.

P3 How have polar bears changed?

Scientists think that polar bears first appeared around 200 000 years ago. They suggest that a few brown bears became isolated from their group by glaciers (ice flows). Over a long time, the bears' fur colour became paler. They also developed other features to help them to live and catch food on the ice.

M1 How did the bears adapt to their environment?

Individuals of every organism on the planet (with a few exceptions) are different. Scientists call this genetic variation. One of the reasons they're different is that they have different sets of genes. In modern-day brown bears, scientists have observed that the colour of the bears' coats shows a wide variation of shades from dark to pale.

In their new environment (the Arctic ice), the bears with the palest coats would have been better camouflaged. This would have made them better hunters of seals for food. These bears would have survived while others would have starved. They would have passed on their genes for their pale coat colour. This is called **genetic adaptation**.

Natural selection

Naturalist Charles Darwin, in his 1859 book *The Origin of Species*, wrote that a 'struggle for existence' was going on all the time in a population of organisms. Because of genetic variation, some

...genetic variation

individuals have characteristics that give them a better chance of survival than others. For animals, this would be those best able to catch food and to escape predators, and those most resistant to disease. These survivors then pass on these characteristics. Darwin called this process **natural selection**.

Darwin, along with another biologist, Alfred Russel Wallace, thought about the possibility of a changing environment. They suggested that, if the environment changed, over time the characteristics of an organism would change. This could eventually lead to the development of a new species. We call this process **evolution**.

What evidence is there for natural selection?

Evidence that natural selection takes place has come from studies of the peppered moth. The wings of this moth are white, 'peppered' with black spots. But sometimes completely black peppered moths occur. This is called the **melanic form** of the moth. A hundred years ago, in the countryside, the melanic form was very rare. But studies in the 1950s suggested that the melanic form was spreading throughout peppered moth populations.

Peppered moths are eaten by birds. Scientists suggested that in rural areas, normal peppered moths resting on trees would be well camouflaged. These trees had organisms called lichens growing on their bark, and were pale in colour. In urban areas, which were becoming more and more polluted in the 1950s, lichens on trees were sensitive to pollution and were killed. Trees became covered in soot. The black melanic moths therefore had a selective advantage and were found in greater numbers.

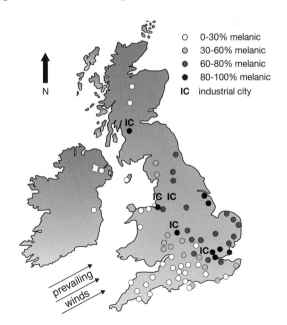

○ 0-30% melanic
◐ 30-60% melanic
● 60-80% melanic
● 80-100% melanic
IC industrial city

The distribution of peppered moths in the 1950s

The melanic peppered moth

🔍 ...evolution and natural selection

The interdependence of organisms

Relationships in danger

Simon is an ecologist working for a conservation organisation. He is studying the interaction between species of animal and plant in British woodlands. He is concerned that, as our environment changes, these relationships will change.

Your assessment criteria:

P3 Describe the interdependence and adaptation of organisms

M2 Explain how organisms within an ecosystem interact over time

P3

Organisms depend on one another

Every organism on the planet interacts with its physical environment. All organisms, however large or small, also interact with, and depend on, other organisms. Ecologists use the term **ecosystem** to describe a habitat and all the organisms that live in it.

The main way that plants, animals and micro-organisms depend on one another is for food. Many animals eat plants; others eat animals that have eaten plants. Most micro-organisms, which include bacteria and fungi, feed on dead or living plants and animals.

Food chains and webs

Ecologists show which organisms feed on each other using types of flow charts called **food chains** and **food webs**. These show how food (and therefore energy) moves from organism to organism.

Feeding relationships between organisms are usually complex. Not many organisms eat just one type of organism. Food webs are therefore more realistic than food chains.

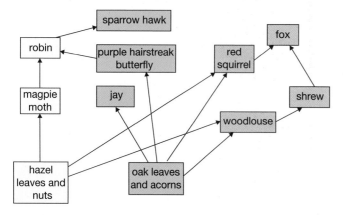

Energy enters the food web as plants photosynthesise. Energy is transferred from organism to organism. Arrows show the direction of flow. Some energy is lost as it goes from one level to the next.

Part of a food web in an oak woodland ecosystem

...food chains ...food webs

Ticks and tapeworms are parasites

The lynx hunts the snowshoe hare

P3

Different types of feeding relationship

A polar bear is a **predator** that hunts, kills and eats seals. The seal is called the **prey**. An animal can act as both predator and prey: a seal eats fish.

A **parasite** lives on, or inside, the body of another organism, which is called the **host**. The parasite takes food from its host. Some parasites kill their host. Others cause disease or weaken the host.

- The parasite that causes malaria kills at least one million people per year.
- All viruses are parasites. HPV is a virus that is thought to cause cervical cancer.
- The tapeworm lives in the human gut. It can cause ill health but rarely death.

M2

Predator–prey relationships

The relationship between predator and prey is one of high interdependence. One of the best known studies of predator–prey relationship is that between the Canadian lynx and snowshoe hare. Fur trappers in Canada kept detailed records of the numbers of the animal skins traded. In some parts of North America, the snowshoe hare is the only prey of the lynx, and the lynx is one of the main predators of the hare.

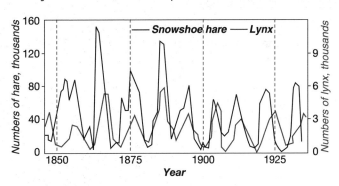

The numbers of snowshoe hares and lynxes showed cycles over a 90-year period

- As numbers of hares increase, there will be more prey.
- The number of lynxes increases.
- More prey will be eaten, so numbers of hares will fall.
- Food for the lynxes becomes scarce, so their numbers will fall.
- The numbers of hares will increase, and the cycle will continue.

Think about

What would happen to a predator–prey cycle if the prey became diseased and died?

Research

Large predators once lived in Britain. The bear and the lynx were probably hunted to extinction in Roman times, and the wolf in the 1600s. Conservationists are thinking about re-introducing these animals in some wild areas of Britain. Find out more about these proposals, and list reasons for and against them.

 ...predator–prey relationship

Human impact on the environment

Your assessment criteria:

P4 Carry out an investigation into the impact of human activity on an environment

> Practical

M3 Describe how to measure the effect of human activity on an environment

D2 Explain how the environmental effect of human activity might be minimised in the future

A world in balance

Sam works for the Environment Agency. She is one of the many scientists across all disciplines – biologists, chemists and physicists – who are investigating our changing environment. Sam believes that if we don't do things to try to prevent the climate changes she's observing, it could have serious effects on all the organisms on the planet.

P4

How do humans affect the environment?

The environment is changing rapidly. We are building more homes and growing more crops for the increasing human population. We are travelling more. We want improved products and technology. More and more energy is needed to fuel all this. As we burn more fossil fuels for energy, **pollutant gases** are produced.

Cars produce carbon dioxide, carbon monoxide, nitrogen oxides, hydrocarbons and tiny particles of carbon called particulate matter

...car exhausts

Trees in this forest have been killed by acid rain

Acidification

Many of the pollutant gases, such as carbon dioxide, nitrogen oxides and sulphur dioxide, dissolve in water to form acids. The gases dissolve in rain, which then falls as acid rain. Evidence from dying trees has shown that acid rain is affecting many parts of Europe.

Carbon dioxide is very soluble in water. The gas dissolves easily in the sea, lakes and ponds. Measurements have shown that these are becoming more acidic. Many organisms will die, and food chains will be seriously affected.

Coral reefs are dying because seawater has become more acidic and warmer. Coral reefs are an important part of the world's food webs

Climate change

Carbon dioxide is also a **greenhouse gas**. It's present in very low concentrations in the air (0.04%), but this level has been increasing rapidly in the last 50 years. The huge quantities of carbon dioxide that are being produced by human activities surround the Earth like a blanket. This reduces the amount of heat that the Earth radiates into space. We call this the greenhouse effect (see Unit 1).

As a result, the Earth is becoming warmer. This rise in average global temperature is causing the world's climate to become more unstable. We can expect periods of very wet or very dry weather, with the likelihood of severe storms. Ice in the polar regions and in mountain glaciers is melting, causing sea levels to rise. The map on the next page shows how parts of the UK and mainland Europe would be affected by a 2 m rise in sea levels.

Edges of the largest glacier in South America, the Upsala Glacier, are shrinking rapidly

 ...greenhouse effect

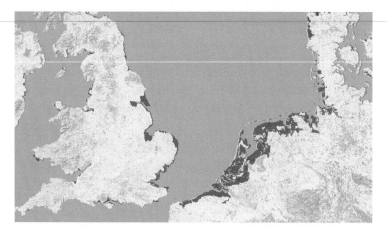

If sea levels rise by 2 m, the red areas would be under water

Know more

Sea levels rose by between 10 and 20 cm in the 20th century. The rate of rise is increasing.

M3

Monitoring levels of pollution

The Environment Agency is one of a number of organisations in the UK that continually monitors the levels of pollutants in the air, in water and on the land.

Many local authorities and energy companies also monitor pollution and publish data on their websites.

Sam says:

'I measure levels of chemical pollutants in rivers and streams. I also look for other characteristics of the water, such as pH and dissolved oxygen concentration.'

Monitoring the effects of climate change

Meteorologists (climate scientists) and other scientists monitor the world's climate. They also monitor the world's ice caps and sea levels.

Biologists monitor the effects that climate change is having on living organisms. As it becomes warmer in Britain, spring is getting earlier. The **distribution** of organisms is changing – many move northwards as the temperatures rise.

Research

Find out about how lichens, which are a mixture of a fungus and an alga, can be good indicators of levels of pollution in the air.

Birds are nesting 1 to 3 weeks earlier than they did 25 years ago

The comma butterfly is spreading northwards

Discuss

Life cycles of plants and animals have to be carefully synchronised so that food and shelter are available at the right time of year. What would happen to butterfly caterpillars if they hatched before trees were in leaf? What effect would this then have on birds that ate the caterpillars?

...climate change

D2

What can we do?

Greenhouse gas emissions in the UK have fallen by about a fifth since 1990, but we need to do much more. Some suggestions are:

- use more energy-efficient processes in industry
- clean up emissions before they're released into the atmosphere
- use alternative energy technologies to generate electricity
- use more energy-efficient vehicles, and vehicles that run on biofuels.

It's also important that we reduce energy use around the home by turning off electrical appliances and insulating our houses. Many people also make use of alternative energy. Grants are available to install solar power and wind turbines to supply our homes with energy.

An environmentally friendly housing development in Hackbridge, Surrey

Carbon-neutral fuels

When a biofuel crop is grown, it absorbs carbon dioxide from the air. This is released as the plant is burnt for fuel. So there is no overall, or 'net', output of carbon dioxide. The biofuel is said to be a **carbon-neutral fuel**. This is in contrast to the burning of fossil fuels. Here, carbon dioxide that was locked away millions of years ago is released today. This has a huge impact on the world's carbon balance.

Agriculture and the environment

Looking at land use

As the population of the UK grew in the second half of the 20th century, there became an ever greater demand for food and new housing. Areas of land were converted into farmland, while other areas were built on.

Know more

> Over 600 different species of plant, 1500 species of insect, 65 species of birds and 20 species of mammals have been recorded, at some time, living in British hedgerows.

P4

Farming in the UK

After the Second World War, farmers were encouraged to increase their yields of crops. Hedgerows were removed so that farmers could grow a single crop in huge fields. This is called **monoculture**. It is much easier to move agricultural machinery around larger fields.

Hedgerows are rich in wildlife, so their removal has a serious effect on the numbers and types of plant and animal species living in the area. The range of different organisms in a habitat is known as **biodiversity**. A single plant crop in a large field will lead to lower biodiversity than smaller fields surrounded by hedgerows.

Using chemicals to improve production

To obtain improved yields of crops, farmers use **fertilisers** and **pesticides**. Fertilisers provide extra minerals for plant crops and increase their yields. But chemicals from the fertilisers can get washed into lakes and ponds. These become enriched with nutrients. This process is called **eutrophication**. Microscopic algae grow in huge numbers and use up the oxygen in the water. All the organisms in the water will eventually die.

Pesticides may be **insecticides**, **fungicides** or **herbicides**. Insecticides kill insects that are pests of the crops. Herbicides kill weeds that compete with the crops for light and minerals. Fungicides kill fungi that are parasites of crop plants. Many pesticides previously in use have now been found to be toxic to other organisms.

Our traditional 'patchwork quilt' of fields surrounded by hedgerows has been replaced by monoculture

Know more

> An insecticide called DDT was widely used for many years. In the 1960s, researchers found that it accumulated in aquatic food chains to toxic levels. DDT was banned in the UK in 1984. But it's very slow to break down and is still present in food chains – even in you.

...biological indicator species

Know more

Forests in tropical regions, called rainforests, have huge biodiversity. Scientists estimate that about half the world's rainforests – that's around 5 million km² – have already been cleared.

Deforestation

P4

Across the world, large areas of forests are being chopped down or burnt to make way for agriculture, to provide timber for building, and to search for important mineral resources.

Deforestation contributes to climate change, because it leads to increased levels of carbon dioxide in the atmosphere. The removal of trees decreases biodiversity, as it destroys habitats. It can also lead to **soil erosion**. Without the protective trees and their roots, heavy rains can wash away soil, leading to landslides.

Monitoring species diversity

M3

Scientists called ecologists carry out surveys of animal and plant life. They record the species that are found in different habitats, and how abundant each species is. They monitor any changes.

Monitoring chemicals in water

Environmental scientists use chemical and instrumental techniques to monitor the levels of fertilisers, pesticides and other harmful chemicals that get washed into water. They also look for the presence of certain organisms in water, called **biological indicators**, to assess how polluted the water is.

Caddis fly larvae are good biological indicators. They need clean water to live

What can we do to halt or reverse change?

D2

In recent years, governments, conservationists, farmers and other organisations have been attempting to conserve habitats and species.

- Some farmers use organic farming methods, without harmful chemicals, to control pests.

- Wildlife Trusts work locally to create **conservation areas.**

- Projects are underway to recreate ancient forests and to reflood wetlands.

- 'Skylark strips' and 'beetle banks' have been created in fields of monoculture crops to provide new habitats for insects; these encourage predators that will also feed on insect pests.

Research

A technique called 'bioremediation' uses living organisms to remove toxic chemicals, such as heavy metals and petroleum products, from land or water. Research a bioremediation project that has been successful.

Discuss

Carbon-neutral crops are grown to produce biofuel. They require huge areas of land. Discuss the environmental value of these.

Research

We can recycle paper products, metals and plastics to conserve natural resources. This may reduce exploration for, and mining of, natural resources. Research and evaluate a recycling scheme in your local area.

 ...conservation of species

Micro-organisms and disease

Your assessment criteria:

P5 Describe the effect of different internal and external factors on human health

M4 Explain how selected medical, social and inherited factors disrupt body systems to cause ill health

D3 Describe the social issues which arise as a result of the selected medical, social and inherited factors and the illnesses they cause

Measles, mumps and rubella

Measles, mumps and rubella used to be common childhood illnesses. The symptoms of these were often quite mild. Some people didn't show any symptoms, or felt only slightly unwell. But sometimes there were complications, and the illness became much more serious. Now children are immunised against these diseases with the MMR vaccine.

P5 Virus infections

Most illnesses in humans are caused by micro-organisms — either bacteria or viruses — that infect the body. Measles, mumps and rubella are diseases that are caused by viruses.

Some **viruses**, such as the common cold virus, cause only minor diseases. We soon get over them. But others are much more serious. AIDS is caused by the human immunodeficiency virus (HIV). The virus attacks cells in the body's immune system. HIV now kills more women of reproductive age worldwide than any other disease.

Know more

Symptoms of measles, mumps and rubella are:

- *measles – a fever, a runny nose and a cough, and a blotchy red rash*
- *mumps – the salivary glands swell up*
- *rubella – a fever, and a rash on the face and neck.*

M4 Congenital rubella syndrome

Rubella is only a very mild disease. But the rubella virus can have very serious effects on an unborn baby. If rubella is caught by a woman in the first eight to ten weeks of pregnancy, it may result in the baby being born with serious defects. The heart and brain can be affected, along with cataracts on the eyes and deafness. Up to 90% of babies exposed to rubella in the womb are born with one or more of these problems.

Long-term virus infections

The problem with some virus infections is that the viruses do not leave our bodies. If you've had a cold sore, after the infection has cleared up, the virus lies dormant in your nerves. If you're under stress, or spend too long in the Sun, the cold sore returns.

Know more

A type of 'flu' (caused by an influenza virus) killed around 50 million people worldwide in 1918–1919.

...immunisation

An HPV vaccination programme started in the UK in September 2008 to help protect girls from cervical cancer

M4

When viruses enter our cells, some can affect the **DNA** and lead to cancer. Nearly all women who develop cervical cancer (cancer of the neck of the womb, or cervix) show signs of having had an infection of the human papilloma virus (HPV).

Vaccines and immunisation

When you get a mild illness, such as a cold, you usually get better on your own. But with many diseases, such as measles, mumps and rubella, it's better to prevent them.

The MMR vaccine is made using live measles, mumps and rubella viruses. But these have been weakened so that they won't cause disease. That's how many vaccines are produced.

When the vaccine is injected, the body produces proteins called antibodies. The antibodies help the white blood cells to destroy the weakened viruses.

After the vaccination, cells called 'memory cells' stay in the body. These will produce antibodies very quickly if the person is ever exposed to the viruses. The person has been **immunised** against the infection, and will be immune for some years.

Think about

Some people say that as the incidence of a disease falls, the risk of contracting it becomes small. So why vaccinate?

Research

In 2000, two studies (now known to be flawed scientifically) suggested that the MMR vaccine might be responsible for the increase of a complex condition called autism in children. For several years, some parents refused to allow their children to have the MMR vaccine. What effect did this have on the number of cases of measles, mumps and rubella?

D3

How bacteria and viruses are transmitted

Disease-causing micro-organisms are transmitted from person to person by direct physical or sexual contact, or by droplet infection (when an infected person sneezes, coughs or breathes into the air).

In an **epidemic**, a disease spreads widely throughout a community. In a **pandemic**, a disease spreads across countries or even the world. We have to take measures to prevent such a spread.

The fight against disease

Antibodies produced by the body are always specific to one type of bacterium or virus. So different vaccines are required for different diseases.

In the UK, we have a vaccination programme to help protect our children against serious diseases they may encounter. But it's difficult to vaccinate everyone. And some parents are discouraged from having their child vaccinated. You normally get just slight discomfort when you've had a vaccination. But there's always a *slight* risk from any medical procedure, however safe it's thought to be. A very, very small number of children have a severe reaction when a vaccination is given to them.

...vaccination

Carcinogens

Your assessment criteria:

P5 Describe the effect of different internal and external factors on human health

M4 Explain how selected medical, social and inherited factors disrupt body systems to cause ill health

D3 Describe the social issues which arise as a result of the selected medical, social and inherited factors and the illnesses they cause

Getting a tan

Jodie has been going regularly to a tanning salon and spending long periods on a sunbed. One day she notices a dark freckle on her arm. She is worried and visits her doctor. The doctor doesn't think the freckle is skin cancer, but she warns her that people with her pale skin colour should not use sunbeds and should avoid sunbathing.

P5

Radiation as a cause of cancer

Ultraviolet radiation (see Unit 2) from the Sun or from sunbeds can damage skin cells. This ages our skin and can lead to a type of skin cancer called a **melanoma**. These occur mostly in fair-skinned people.

Any agent that increases the risk of cancer is called a **carcinogen**. Some carcinogens are physical agents, like radiation. Ionising radiations, such as those produced by radioactive materials (see Unit 2) have been shown to cause cancer. In the UK many of us are exposed to natural radiation all the time. Radon is a radioactive gas that seeps from the ground. It's thought to be the second most important cause of lung cancer.

Chemical carcinogens

Many different chemicals are known to cause cancer. Others are suspected of increasing the risk. Some workers in the UK are exposed to chemical carcinogens. Several examples are given in the table.

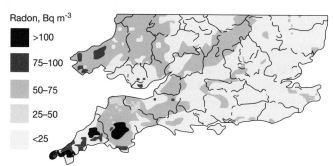

Radon, Bq m^{-3}

■	>100
■	75–100
■	50–75
□	25–50
□	<25

High concentrations of radon are found in the south-west of Britain

Chemical carcinogen	Type of cancer	People most at risk
asbestos	lung	demolition workers
benzene	bone marrow	painters; workers in the petrochemical industry
hair dyes	bladder	hairdressers
methanal (formaldehyde)	nose	hospital lab workers; manufacturers of wood products, e.g. MDF
vinyl chloride	liver	plastics workers

...carcinogen

P5

Smoking and cancer

Cigarette smoke is the most common way in which people become exposed to chemical carcinogens. It contains many chemicals that are known to cause cancer. These include 1,2-benzopyrene, benzene, methanal and arsenic.

Studies in Europe, the USA and Japan have shown that nine out of ten people who get lung cancer are smokers. Research also suggests that people who start smoking as teenagers are ten times more likely to develop lung cancer than those who have never smoked.

Smoking has also been linked with other types of cancer, including the mouth, oesophagus, kidney and bladder.

Know more

It would be too costly and time-consuming to investigate whether many thousands of suspected chemical carcinogens do cause cancer in people. Scientists therefore often use a test called the Ames test. They look for mutations when a bacterium is exposed to the chemical.

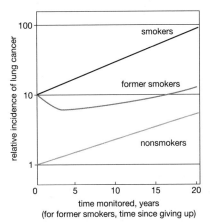

Lung cancer rates in people who have never smoked, in those who smoke, and those who have given up smoking

What happens when we develop cancer?

M4

When a cell becomes cancerous, it starts to divide uncontrollably. Many new cancer cells are produced. The cancer cells often form a clump of cells called a tumour. Sometimes some of these cancerous cells spread to other locations in the body. Here, they form new tumours. There comes a point when surgery cannot remove all the tumours from a person's body.

It's a change to the DNA in a cell that causes it to become cancerous. This is called a **mutation**. Carcinogens cause harmful mutations. Some mutations are caused when cells become infected by certain micro-organisms. Not all mutations cause cancer.

Discuss

The chemicals in cigarette smoke also cause heart disease, and a lung disease called emphysema. Where there's a hospital waiting list, some people say that non-smokers should be given treatment first. It's even been suggested that smokers should not be treated at all, unless they change their smoking habit. What do you think?

Do our lifestyles cause cancer?

D3

Some factors in our lives undoubtedly contribute to cancer and other illnesses. Smoking is a significant cause of cancer, both in people who smoke, and those who inhale other people's smoke. In the industrial world, diets are often high in fat and low in fibre. This is linked to high rates of cancer of the large intestine (colon and rectum). Some scientists think that exposure to microwave radiation from mobiles may increase the risk of brain tumours.

Social impact of cancer

Cancer and its treatment can affect all aspects of a sufferer's life. Individuals who have cancer, and their families, need practical and educational, emotional and sometimes financial support. Being able to share experiences with someone in the same position, or someone who has survived the disease, can often help patients to cope.

The use and misuse of drugs

Your assessment criteria:

P5 Describe the effect of different internal and external factors on human health

M4 Explain how selected medical, social and inherited factors disrupt body systems to cause ill health

D3 Describe the social issues which arise as a result of the selected medical, social and inherited factors and the illnesses they cause

Unfit to drive

Jason is a police officer. He suspects a driver to be under the influence of drink or drugs. He stops the car and carries out a roadside breath test on the driver.

P5

Alcohol

Alcohol affects how our brain functions. It is a **sedative**, or depressant. It lowers a person's awareness of what's going on. Even low amounts of alcohol increase a driver's response time. This could have fatal consequences on the road.

Higher amounts of alcohol can inhibit the part of the brain that controls breathing, and even cause death.

Other misused drugs

There's no roadside test for so-called 'recreational' drugs, but Jason checks the person's eyes, balance, walking and co-ordination. Like alcohol, these drugs impair concentration and slow reaction times, and the driver can often misjudge speed.

Recreational drugs act on the brain and other parts of the nervous system to affect mood, perception and consciousness.

Drug	Effect
amphetamines, e.g. ecstasy, speed	stimulants
cannabis	changes a person's mood; produces feelings of closeness to people and experiences of increased intensity
cocaine	stimulant
LSD	alters perception and produces hallucinations
opiates, e.g. heroin, morphine, codeine	have a painkilling effect and produce a feeling of euphoria (happiness)

Know more

Tranquillisers and barbiturates are also depressant drugs.

Know more

Caffeine (in coffee, tea and some high-energy drinks) and nicotine (in cigarettes) are also stimulants.

Know more

Some drugs affect the way our body holds on to water and how much urine we produce.

Alcohol and caffeine will increase the amount of urine produced.

Ecstasy has the opposite effect on water balance. It decreases the amount of urine produced.

...drugs and alcohol

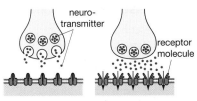

A neurotransmitter is released from the first nerve. It moves across the gap between the nerves. As it reaches the next nerve, it locks into a receptor. The second nerve is stimulated.

Nerve impulse transmission

The effect of drugs on the nervous system

M4

The messages that nerves carry around the body are called nerve impulses. A nerve impulse is carried from one nerve to another by a chemical called a **neurotransmitter**. Drugs affect different types of neurotransmitter in the brain, or the receptor molecules they lock on to. This is how they produce their effects.

Drugs also affect other systems in the body, often producing unwanted effects. Our body has control systems that keep its internal environment constant. Many drugs affect this control.

Research

The use of drugs affects many organs within the human body. Research why and how the liver may be one of the major organs affected by long-term drug use.

Many drugs are addictive

D3

Many drugs can have an impact on long-term behaviour. People can quickly become dependent on some drugs. The person develops a need for the drug. This need might be psychological, when the person finds it difficult to get through the day without the drug. Or it may be **physiological**, when the person's body can no longer function without the drug.

Treatment and rehabilitation – getting the person back to full health and a normal life without the drug – are important.

Wider issues arising from drug misuse

The misuse of drugs presents a challenge to the stability of the user's family. Behavioural changes as a result of the drug use can lead to crime. Or the user may turn to crime to obtain money to fuel the drug habit.

Research

Recreational drugs are illegal. They are divided into three classes: A, B and C. Find out which drugs are currently in each class, and what the classes mean. From your knowledge of the drugs' effects, explain why each has been placed in the different classes.

The use of recreational drugs has led to an increase in the transmission of disease. It can also result in unwanted pregnancies. Mental and behavioural problems and accidents can lead to the death of the user.

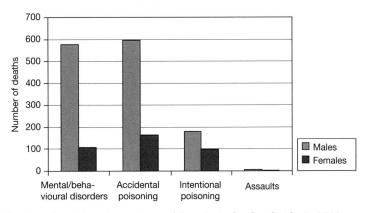

Deaths related directly to misuse of drugs in England and Wales in 2008

Discuss

Some countries, such as the Netherlands, have more liberal laws on drug use than the UK. Some people believe that the laws in this country should be relaxed. They say that more easily available, cheaper drugs would mean less crime. Discuss this issue from a scientific and a social viewpoint.

...drug addiction

Inherited conditions

Your assessment criteria:

P5 Describe the effect of different internal and external factors on human health

M4 Explain how selected medical, social and inherited factors disrupt body systems to cause ill health

D3 Describe the social issues which arise as a result of the selected medical, social and inherited factors and the illnesses they cause

Cystic fibrosis

Jack has a disease called cystic fibrosis. He needs vigorous physiotherapy every few days to keep his lungs working. Fifty years ago, children with cystic fibrosis would have been unlikely to live beyond the age of five. Today, half will live beyond the age of 35. And scientists are working all the time to improve this.

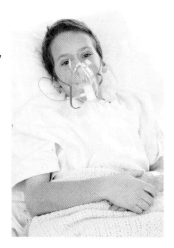

P5

Inherited diseases

Cystic fibrosis is an example of a disease that's inherited, or genetic. Sufferers produce mucus that is thicker and stickier than normal. This mucus blocks Jack's airways in his respiratory system, making it difficult for him to breathe. His respiratory system can get infected easily.

Cystic fibrosis also reduces the secretion of digestive juices from the **pancreas**. This makes it more difficult to digest food.

Type 1 diabetes is another disease that affects the functioning of the pancreas. As well as producing digestive enzymes, the pancreas produces a hormone called insulin. Insulin helps us to regulate our blood glucose (sugar). If a person has type 1 diabetes, the cells in the pancreas that produce insulin have been attacked and damaged by the person's *own* immune system. It's therefore described as an **autoimmune disease**. The person must have regular injections of insulin and also control their diet very carefully.

We know that autoimmune diseases tend to occur in families. But family members over several generations are likely to have *different types* of autoimmune disease.

Scientists now recognise about 150 diseases as being autoimmune diseases. These include:

- Crohn's disease, which affects the digestive system
- multiple sclerosis, which affects the brain and spinal cord
- rheumatoid arthritis, which can affect the whole body.

Know more

The body's control of glucose levels will be looked at in more detail in Unit 10.

Know more

Under normal conditions, an immune response occurs when a micro-organism enters the body. The immune system isn't normally triggered by cells in your own body. It distinguishes between 'self' and 'non-self'.

Know more

Environmental factors are also involved in the development of autoimmune diseases. These include chemicals in the environment, drugs, radiation and micro-organisms.

...single gene disorders

Research

A disease called sickle cell anaemia affects the red blood cells. Find out more about the disease, its effects on the body, and why it's common in African populations.

The genetics of cystic fibrosis

Cystic fibrosis is caused if a certain gene is defective. The gene is located on chromosome 7. It codes for a protein called CFTR. This protein controls the movement of water and ions in and out of certain cells, such as those of the **epithelium** lining the lungs.

One of each of our pairs of chromosomes came from our mother, and one from our father. If a person has one normal gene, the person can still produce CFTR. The person does not have cystic fibrosis but is a **carrier**. They can pass the defective gene on to their children. A person who has two defective genes has cystic fibrosis.

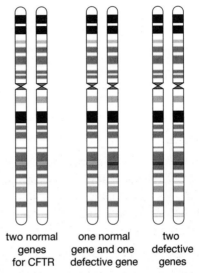

| two normal genes for CFTR | one normal gene and one defective gene | two defective genes |

It takes two defective genes for a person to suffer from cystic fibrosis

How can we help sufferers of cystic fibrosis?

The need for regular **physiotherapy** means that cystic fibrosis creates severe disruption to family life. Jack also has to do regular exercises to improve his health. He takes enzyme tablets to help him to digest his food and has regular visits from his community dietician. He and his family also have counselling to discuss the emotional and psychological effects of the disease.

With our knowledge of genetics, we can sometimes predict the possibility of someone being born with cystic fibrosis. Over two million people in the UK carry the cystic fibrosis gene. That's about 1 in every 25 of the population. A couple who have cystic fibrosis running in both families may decide to have a simple test, using cheek cells or blood cells, to see if they are carriers. They can then make a decision as to whether they have children. If both parents are carriers, there's a 1 in 4 chance of any children born having cystic fibrosis.

Research

Scientists are attempting to use gene therapy to treat cystic fibrosis. They are hoping that one day they may be able to cure the disease by introducing a normal version of the cystic fibrosis gene to the person's epithelial cells. Find out about gene therapy, and how successful it's been so far.

Diet and exercise

Your assessment criteria:

P5 Describe the effect of different internal and external factors on human health

M4 Explain how selected medical, social and inherited factors disrupt body systems to cause ill health

D3 Describe the social issues which arise as a result of the selected medical, social and inherited factors and the illnesses they cause

Healthy diets

Sarah is a community dietician. She gives advice to people on how to stay healthy through a proper diet and an appropriate lifestyle. She also helps to develop nutritional plans for people who need a special diet for different reasons. They may have food allergies or they may want to train for a marathon.

P5

A balanced diet

Sarah is concerned about the diet of many people in her local community. She says:

'Many people don't find the time to eat properly. I explain that our diet should contain types of food called carbohydrates, proteins, lipids (fats), vitamins and minerals. And it's important that these must be in the correct proportions to provide a balanced diet.'

The food we eat provides the raw materials for growth, and for the normal functioning and repair of cells and body tissues. It also provides the energy we need to stay alive and be active.

Food nutrient	Role of nutrient in the body	Sources of the nutrient
carbohydrates	provide energy	bread, pasta, potatoes, rice
proteins	for growth and development, and repair	animal proteins: meat, fish, eggs, cheese non-animal proteins: nuts, beans, cereals, Mycoprotein
lipids (fats)	for growth and development, e.g. of nervous tissue	animal fats, vegetable oils, butter and margarine

Part of Sarah's information leaflet on a balanced diet

...diet and health

P5

Carbohydrates, proteins and lipids (fats) are required in relatively large amounts in a balanced diet. Vitamins and minerals are required in much lower amounts, but are needed for healthy growth and development. Some are necessary for enzymes to function and for us to obtain energy from our food.

Sarah continues:

'I encourage people to increase the amount of fish oil in their diet for good health, and to use plant oils rather than solid animal fats to reduce the risk of heart disease. I also suggest that they eat more wholegrain sources of carbohydrate, which release glucose slowly into the blood.'

M4

Balancing energy requirements

We all need a certain amount of energy just to live. This is called our **basal metabolic rate**, or **BMR**. Depending on age and gender, our BMR is 45 to 70% of our total energy expenditure. So the amount of energy we need to take in from our food depends on our age and gender. Also, the more active a lifestyle a person has, the more energy they need.

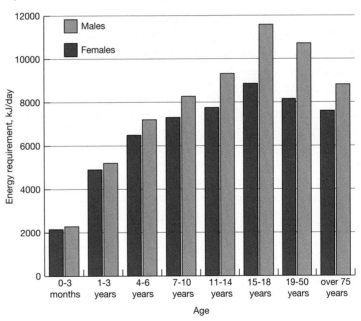

The estimated average energy requirements of people of different ages

It's important to balance the energy in the foods we eat with the amount of energy we need for our daily activities. In the UK today, many people eat food containing far more energy than they actually need. Our bodies convert excess carbohydrate to fat. This gets stored under our skin and around body organs. This fat can make us overweight or obese.

Activity	Energy requirement, kJ/h
sitting	420
typing rapidly	590
dressing	630
walking, 3 mph	840
jogging, 6 mph	2400

The energy required to do various activities

The importance of exercise

It is easy to eat more than we need if we don't take very much exercise. Exercise isn't only important for our energy balance. It's also important for our overall health. It keeps our heart and other muscles in our body healthy. And chemicals released during exercise also affect our mood and give us a sense of well-being.

Body mass index

Sarah continues to talk about her work.

'People don't often realise how their need for energy depends on their body size. I can show people how to calculate their body mass index (BMI), to find out if they are the ideal weight for their height, or over- or underweight.'

$$BMI = \frac{weight,\ kg}{(height,\ m)^2}$$

'I have to explain that the information from a BMI can sometimes be misleading. It's inappropriate for pregnant women, for instance. And a BMI might suggest that a very fit person, with a high proportion of muscle, is obese!'

Consequences of the wrong diet

Too much food, or a diet rich in animal fats, can lead to serious conditions such as heart disease.

- If a person is overweight, their heart has to work more to deliver oxygen to body cells. This puts strain on the heart.

- Excess fat in the diet can lead to fatty deposits that build up in the arteries (blood vessels). The arteries are made narrower and can become more rigid. This increases blood pressure. High blood pressure can lead to a stroke if a blood vessel bursts, flooding the brain.

- If fatty deposits build up in the coronary arteries which supply blood to the heart, then a blood clot could form. This can cause a heart attack.

Diabetes (type 2) and cancer are also more likely if a person is overweight.

Think about

Sarah advises women who are pregnant and breastfeeding about their extra energy requirements.
The recommended extra intake of energy during the last three months of pregnancy is 80 kJ per day. Someone who's breastfeeding needs an extra 2100 kJ per day. Why is this extra energy required?

The BMI calculator shows you quickly if you are overweight, within a normal range, or overweight

...BMI calculator

Lifestyle factors

D3

Sarah appreciates, from talking with people in her community, why they don't always have healthy lifestyles. Some of their comments are shown here.

Discuss

Discuss what these people say, and suggest what they might try to make their diet or lifestyle more healthy.

It's quicker to cook a 'ready meal' than prepare more healthy food.

Fresh fruit and vegetables are more expensive than fast food.

I'm so busy that I can't find the time to exercise.

I can't get my children to eat vegetables.

Research

In some parts of the world, diets have insufficient protein owing to ecological and social factors. Diseases called kwashiorkor and marasmus occur. Find out more about these diseases.

Anorexia nervosa

Some people that Sarah works with have a different problem. They have a condition called **anorexia nervosa**.

Because of the society we live in, many people feel under pressure to be thin. But some have a mistaken perception of their body size. They may imagine themselves to be overweight when actually they're not. They reduce their food intake to very low levels.

With insufficient protein in their diet, muscles are broken down to provide energy and the body wastes away. A lack of glucose reduces mental function. Anorexia that's untreated is usually fatal.

Sarah says:

'I help patients to gradually establish healthy eating habits to restore their body weight. It is also important to try to tackle the reasons that lead people into this condition.'

Research

One of Sarah's patients has Crohn's disease. Crohn's disease is an autoimmune disease affecting the gut. Find out more about this and how diet is important in helping to control the condition.

Control mechanisms

Running the marathon

Eve is running a marathon. Her big day has arrived. She must now run over 26 miles. While she is running, mechanisms in her body will try to keep internal conditions in her body as constant as possible.

P6 — Respiration releases heat

As Eve is running, the rate at which she respires increases. Her body cells break down glucose to release the energy she needs to run. As the energy in glucose is transformed into the energy required for her to run, heat energy is also produced.

The heat energy released by Eve's body cells will be carried around her body by her blood. Eve's normal body temperature is 37 °C, and it must remain close to this temperature at all times. The organs such as the brain are very sensitive to changes in temperature.

Eve begins to go red and sweats. These are effects of **thermoregulation** – the regulation of body temperature.

Vasodilation

Blood vessels in Eve's skin, called arterioles, widen. This is called **vasodilation**. More blood will then flow through the narrow blood capillaries at the surface of Eve's skin. Her skin reddens and heat will be lost.

Sweating

Eve will lose most heat from her body by sweating. Sweat is produced by the sweat glands in her skin. Sweat is mostly water, but contains some salt and other chemicals. As the water on the surface of her skin evaporates, it takes heat with it. This is called **evaporative cooling**.

Know more

Respiration will be looked at in more detail in Unit 10.

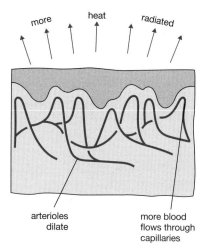

Vasodilation lowers body temperature

Know more

Humans and horses sweat, but many other mammals don't sweat efficiently. Cats and dogs have sweat glands only on the soles of their feet. They pant, and lose heat by evaporative cooling from their lungs.

Homeostasis

P6

The processes by which conditions in the human body are kept constant are together called homeostasis. Thermoregulation is one of these processes. **Homeostasis** in the human body is controlled by two systems.

- The **nervous system** communicates using conducting cells called nerve cells or **neurons**. A message, or nerve impulse, that travels along neurones is an electrical signal.

- The **endocrine system** uses chemical messengers called hormones. These are produced by a series of glands in the body and transported in the blood.

Nervous control: how thermoregulation works

A region of the brain called the **hypothalamus** detects an increase in the temperature of Eve's blood. Messages are also sent to the hypothalamus along nerves, from temperature sensors around her body. The hypothalamus responds and nerve impulses are sent to the skin. Eve's blood vessels dilate, and she begins to sweat.

Know more

As she sweats, Eve loses water from her body. This must be replaced to avoid dehydration. She takes in about three litres of water by drinking regularly during the race.

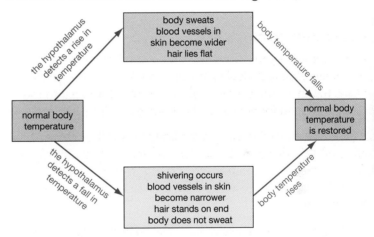

Controlling body temperature

After the race, Eve's body temperature falls. This temperature change is, once more, detected by the hypothalamus. Eve's brain sends nerve impulses to her skeletal muscles. These contract, and she begins to shiver. This releases heat. Eve's body temperature begins to rise.

Endocrine control: controlling glucose in the body

Eve is using up carbohydrate during the race. The stores of carbohydrate in her body become low. She takes in some extra carbohydrate from a sports drinking contain glucose. The level of glucose in her blood increases. The pancreas then releases the hormone insulin, which enables the body cells to take up glucose and use it for respiration.

Know more

Whether exercising or not, the level of glucose in your blood, and its supply to your cells, must be carefully controlled. The body's control of glucose levels will be looked at in detail in Unit 10.

Assessment and Grading criteria

To achieve a pass grade, my portfolio of evidence must show that I can:

Assessment Criteria	Description	✓
P1	Describe how the functioning of organisms relates to the genes in their cells	
P2	Construct simple identification keys to show how variation between species can be classified	
P3	Describe the interdependence and adaptation of organisms	
P4	Carry out an investigation into the impact of human activity on an environment	
P5	Describe the effect of different internal and external factors on human health	
P6	Identify the control mechanisms which enable the human body to maintain optimal health.	

Zygaena butterfly

To achieve a merit grade, my portfolio of evidence must show that I can:

Assessment Criteria	Description	✓
M1	Describe how variation within a species brings about evolutionary change	
M2	Explain how organisms within an ecosystem interact over time	
M3	Describe how to measure the effect of human activity on an environment	
M4	Explain how selected medical, social and inherited factors disrupt body systems to cause ill health.	

To achieve a distinction grade, my portfolio of evidence must show that I can:

Assessment Criteria	Description	✓
D1	Explain how genes control variation within a species using a simple coded message	
D2	Explain how the environmental effect of human activity might be minimised in the future	
D3	Describe the social issues which arise as a result of the selected medical, social and inherited factors and the illnesses they cause.	

Unit 4 **Applications of Chemical Substances**

LO

Be able to investigate exothermic and endothermic reactions

- Self heating coffee relies on an exothermic reaction to give out the heat and warm the coffee

- Exothermic combustion reactions heat our homes and fuel our vehicles

- Photosynthesis is an endothermic reaction because it needs light energy from the sun to work

LO

Be able to investigate chemical substances with different types of bonding

- Strong covalent bonds make diamond the hardest natural substance we know

- Some crystal shapes can be explained by their ionic bonds

- Conducting electricity depends on the type of bonding present

LO

Know about specialised materials and their applications

- Nanoparticles are measured in billionths of a metre

- Nanochemistry could help us make a computer the size of a sugar lump

- The Kevlar lining in a Bladerunner hoodie makes it bullet and knife proof

- Lenses in glasses that darken in sunlight are smart materials

LO

Be able to investigate organic compounds that are used in society today

- The petrochemical industry makes many everyday essential materials like detergents, medicines and dyes

- LPG fuel is a mix of the organic compounds propane and butane

- Designer plastics can be made to order

- PVC can be rigid like drainpipes or soft like clingfilm; it all depends on what's added

- We know that our ancestors have been brewing beer for at least 12 000 years

Covalent bonding

Your assessment criteria:

P1 Carry out experiments to identify compounds with different bonding types

Practical

M1 Describe the properties of chemical substances with different types of bonds

D1 Explain why chemical substances with different bonds have different properties

The biggest diamond

The Golden Jubilee diamond is thought to be the biggest cut diamond in the world. South African miners discovered it in 1985. It is worth up to £9 million. Diamonds are made from carbon. So is soot. The difference lies in the bonding. The strong chemical bonds between the carbon atoms in diamond make it very hard, sparkly and very valuable.

P1 Making molecules

There are different types of chemical bond. The chemical bonds in diamonds are **covalent bonds**. The outer electrons of the atoms are rearranged when atoms join with covalent bonds. Some of the electrons are shared between the two atoms. This bonds the atoms strongly together. Covalent bonds are very strong bonds.

In a hydrogen molecule, each hydrogen atom shares its electron with the other atom. This is called a **single covalent bond**. By sharing their electrons, each hydrogen atom has a full outer shell (see Unit 1). This makes it more stable.

Covalent bonding in a hydrogen molecule, H_2

single covalent bond

We use dots and crosses to show how the electrons from each atom are arranged

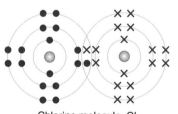

Chlorine molecule, Cl_2

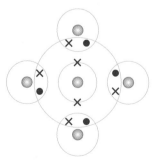

Methane molecule, CH_4

...covalent bonds

Know more

*Nitrogen (N_2) is the gas in crisp bags. It helps keep crisps fresh. Nitrogen molecules have a **triple covalent bond**. Each nitrogen atom shares three electrons with the other.*

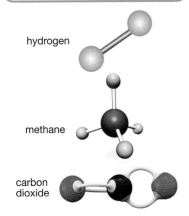

hydrogen

methane

carbon dioxide

Molecular models. Note the representation of double bonds in carbon dioxide

Think about

An electron is an electron. Does where it came from make a difference to the compound?

Discuss

Can you explain why diamonds are the hardest natural substance we know?

Can you explain why diamonds take thousands of years to form?

P1

Some molecules have **double covalent bonds**. Oxygen atoms each share two electrons with the other.

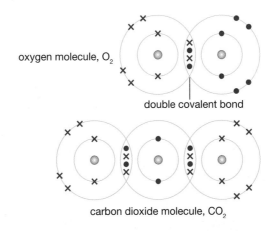

oxygen molecule, O_2

double covalent bond

carbon dioxide molecule, CO_2

M1

Dative covalent bonds

If both the shared electrons come from the same atom, it is a special type of covalent bond called a **dative covalent bond**.

Carbon monoxide is poisonous. It makes a very stable dative covalent bond with the haemoglobin in your blood. This poisons it.

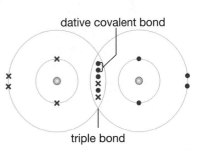

dative covalent bond

triple bond

Dative covalent bond in carbon monoxide, CO

D1

Carbon is a special case

A carbon atom has four electrons in its outer shell. It can make four single covalent bonds. This structure of bonded atoms can go on and on. In three dimensions, it makes a strong **tetrahedral structure**. This is diamond.

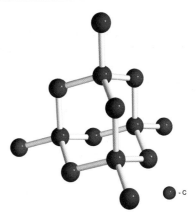

● - C

Diamond structure

🔍 ...tetrahedral structure

Ionic bonding

Do you want salt on that?

We have been adding salt to our food for hundreds of years. It preserves food and gives that favourite salty taste. The salt we use is sodium chloride. This contains the group 1 element sodium and the group 7 element chlorine. They are chemically bonded together with ionic bonds.

Your assessment criteria:

P1 Carry out experiments to identify compounds with different bonding types

Practical

M1 Describe the properties of chemical substances with different types of bonds

D1 Explain why chemical substances with different bonds have different properties

P1 | A reactive metal and a green gas

Sodium is a very reactive group 1 metal. Its atoms have one electron in their outer shell.

- The numbers 2, 8, 1 show the numbers of electrons in each shell.

Chlorine is a very reactive green gas. Its atoms have seven electrons in their outer shell.

- The numbers of electrons in each shell are 2, 8, 7.

When they bond together, sodium's one outer electron is transferred to chlorine. This gives both particles a full outer shell of electrons. It makes them very stable. The new particles do not have the right number of electrons to be atoms. They are called **ions**.

- Losing an electron gives the **sodium ion** a **positive charge**. We write it as Na^+.

- The **chloride ion** has gained an electron and has a negative charge. We write a chloride ion as Cl^-.

These opposite charges attract one another strongly by **electrostatic attraction**. Ionic bonds are strong bonds.

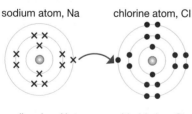

The formula of sodium chloride is NaCl

Fries with sodium chloride, commonly known to us salt

Know more

Electrons are negatively charged particles. Protons in the atom's nucleus are positively charged. Atoms have the same number of protons and electrons. The charges balance out. Losing or gaining electrons gives ions their charges.

Know more

It is a chlorine atom but a chloride ion.

 ...ions

magnesium ion, Mg²⁺   oxide ion, O²⁻

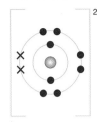

In the ionic compound magnesium oxide, two electrons are transferred from magnesium to oxygen

The word 'salt' to a chemist means a whole family of **ionic compounds**. Sodium chloride is just one of them.

P1

Does an NaCl 'molecule' exist?

M1

The answer is 'No'. A positive sodium ion can surround itself with six chloride ions. Likewise, a chloride ion can surround itself with six sodium ions. It makes a **lattice**. Sodium chloride exists as a **giant ionic structure**.

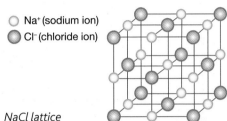

○ Na⁺ (sodium ion)
● Cl⁻ (chloride ion)

NaCl lattice

Sodium chloride crystals

> **Think about**
>
> *Look at the sodium chloride lattice. Can you explain the shape of the sodium chloride crystals in the photo?*

> **Think about**
>
> *What does the formula NaCl tell us about sodium chloride?*

Metallic bonding

D1

Metals have different bonding. Metal atoms lose some of their outer electrons. They become positive metal ions. The lost electrons move around in the metal. The metal ions are attracted to the electrons, and the electrons are attracted to the metal ions. Metallic bonds are also strong bonds.

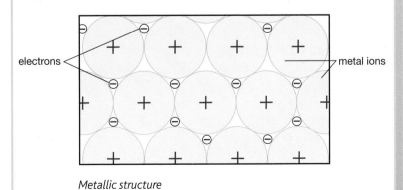

electrons — metal ions

Metallic structure

> **Research**
>
> *How do metals conduct electricity?*

Bonding and properties

What's in our food?

Many of our foods contain **food additives**. They are listed on the label. Labelling must be accurate by law. Additives in food are regularly checked by the Food Standards Agency. Workers identify the additives and find out how much the food contains. Knowing how the atoms are bonded together is the first step in identifying the additive.

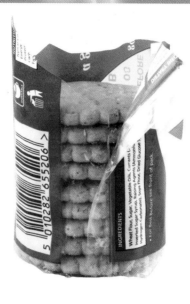

M1 Matching bonding to properties

Workers at the Food Standards Agency know that the type of bonding in an additive affects its properties. Some additives have covalent bonding, others have ionic bonding. They can carry out tests to find which is present.

Appearance

- Many ionic compounds are crystalline solids at room temperature.

- Covalent compounds may be solids, liquids or gases.

Solubility

- Many ionic compounds dissolve in water, but do not dissolve in **organic solvents**.

- Many covalent compounds dissolve in organic solvents, but not in water.

Melting and boiling points

- Ionic compounds have high melting points. This means they are solid at room temperature. Some of the ionic bonds have to break when the solid melts. The bonds are strong, so this needs a lot of energy. The melting point is high.

- Covalent compounds are molecules. The covalent bonds do not have to break when the solid melts. Less energy is needed, so the melting point is lower.

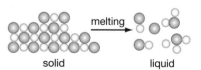

solid → melting → liquid

⬤ chloride ion, Cl⁻

◯ sodium ion, Na⁺

When sodium chloride melts, the giant ionic structure has to break up. Bonds need to be broken

...food additive

M1

Conducting electricity

- Ionic compounds conduct electricity when melted. If they are soluble in water, the solution will also conduct electricity.

- Covalent compounds do not conduct electricity.

Think about

Explain why graphite easily writes on paper.

Graphite is pencil 'lead'

D1

Graphite

Like diamond, **graphite** is made from carbon atoms joined together by covalent bonds. Unlike diamond, each carbon atom makes three, not four, covalent bonds with other carbon atoms. The carbon atoms join up in hexagons forming flat sheets. Spare electrons make a layer between the sheets.

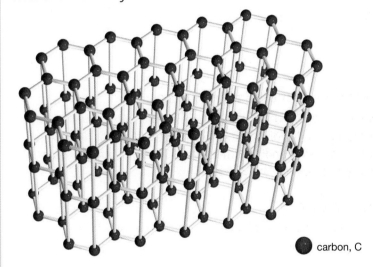

carbon, C

The structure of graphite

Electrical conduction

The following all conduct electricity:

- metals

- melted ionic compounds

- ionic compounds dissolved in water

- graphite.

They all have free ions or free electrons. These are charged particles. They can move and carry the electric charge through the material.

Covalent compounds do not have ions or free electrons to carry an electric charge.

Research

Find out why graphite is the odd one, out of the list here.

Research

Which metals are used to carry electricity to our homes?

Exothermic and endothermic reactions

Self-heating food

The army needs to provide hot convenience food for soldiers away from base. Ideally the meals should need no cooking pans. The solution is self-heating food. Part of the food packaging contains chemicals. The meal is sealed in a separate section. When the chemicals react together, heat is given out. This warms the food.

Your assessment criteria:

P2 Carry out experiments to investigate exothermic and endothermic reactions

> Practical

M2 Explain the temperature changes that occur during exothermic and endothermic reactions

D2 Explain the energy changes that take place during exothermic and endothermic reactions

P2

Hot and cold reactions

When water is added to calcium oxide in a beaker, the beaker gets very hot. Heat is given out. The reaction is:

$$CaO + H_2O \rightarrow Ca(OH)_2$$

Some self-heating food packs use this reaction to warm the food.

This is an **exothermic reaction**. In an exothermic reaction, heat is transferred from the *system* (the reacting chemicals) to the *surroundings* (the beaker, the bench, the air). The temperature of the surroundings rises.

We can show this on an **energy profile diagram**.

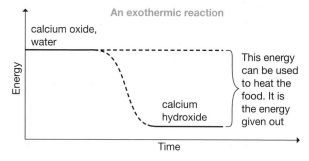

An exothermic reaction

calcium oxide, water

Energy

calcium hydroxide

This energy can be used to heat the food. It is the energy given out

Time

Energy profile diagram: calcium oxide reacting with water

Cold packs are used by athletes to treat sprains. One section of the pack contains water. The other section contains a salt such as ammonium nitrate. When they mix, the ammonium nitrate dissolves. This needs energy so heat is taken in.

This is an **endothermic reaction**. In an endothermic reaction, heat is transferred from the surroundings to the system. The temperature of the surroundings (including the pack) falls. The pack becomes cold.

Self-heating food

Making use of an endothermic reaction

...smart packaging

P2

We can show an endothermic reaction on an energy profile diagram.

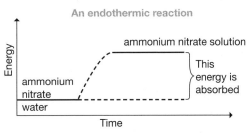

Energy profile diagram: ammonium nitrate dissolving in water

M2

Making and breaking bonds

In a chemical reaction, old bonds are broken and new ones are made. When calcium oxide reacts with water, these bonds are broken:

- the calcium–oxygen bonds in calcium oxide
- the oxygen–hydrogen bonds in water.

New bonds are made in forming calcium hydroxide.

Breaking old bonds needs energy. It is endothermic.

Making new bonds gives out energy. It is exothermic.

It is a balancing act, between the energy needed to break bonds and the energy released in making them. This decides whether the reaction is exothermic or endothermic.

Think about

An acid neutralises an alkali. The reaction is exothermic. The heat given out is called the heat of neutralisation. Which needs more energy: breaking the old bonds or making the new ones?

D2

Showing the energy changes

Bond breaking and bond making can be shown on the energy profile diagram of the reaction.

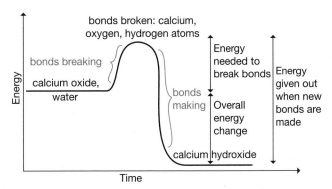

Energy profile diagram: bond breaking and bond making for the reaction between calcium oxide and water

Think about

What information about the reaction does the army need in order to design its self-heating meals?

...calcium oxide

Combustion

Keeping your energy up

Many sports drinks contain glucose. The glucose is a source of energy. Glucose contains carbon, hydrogen and oxygen. In your body cells, the glucose slowly reacts with oxygen. This is respiration. The reaction is exothermic. Energy is given out. This keeps you warm and gives you energy to move.

P2 Using exothermic reactions

Glucose acts as a fuel in our bodies. Fuels like coal, oil and gas also contain carbon and hydrogen. When they burn, they react with oxygen. The reaction is called **combustion**. If there is plenty of oxygen, the products are carbon dioxide and water. The reaction is very exothermic.

When we use these fuels to heat our homes, the fuel and the oxygen are the *system*. Our homes become the *surroundings*. The exothermic reaction transfers heat from the system to the surroundings. The temperature of our home rises.

Different exothermic reactions give out different amounts of heat energy. We can measure the energy given out by some fuels and foods in the lab.

> **Know more**
>
> **Complete combustion** is when carbon dioxide and water are made. There must be plenty of oxygen available.

> **Know more**
>
> If boilers are faulty, they may not take in enough oxygen for the fuel to burn properly. This is called **incomplete combustion**. Carbon monoxide and water are made. Carbon monoxide is a very toxic gas. It is important to have boilers serviced regularly.

🔍 ...respiration

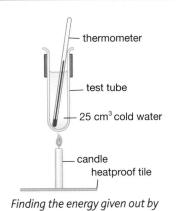

Finding the energy given out by burning candle wax

Measuring candle power

P2

Candle wax also contains carbon and hydrogen. If we measure the mass of the candle before and after it has burnt, we can work out how much candle wax has been burnt. The heat energy given out is used to heat 25 cm^3 water. We can measure the temperature rise of the water.

The set-up shown on the left can be used to measure the heat given out by different fuels, or by foods such as crisps, crackers and corn puffs.

Think about

A Bunsen burner flame can be blue or yellow, depending on the air intake. Which is complete combustion? Which is incomplete combustion?

How much energy was given out when the candle burnt?

M2

We use the formula:

$$\text{energy transferred (joules)} = \text{mass of water (g)} \times 4.2 \times \text{temperature change (°C)}$$

If the mass of water is 25 g and the temperature rise is 35 °C, then:

$$\text{energy transferred} = 25 \times 4.2 \times 35 = 3675 \text{ joules}$$

Where did the energy come from?

D2

An energy profile diagram helps to show us what is happening.

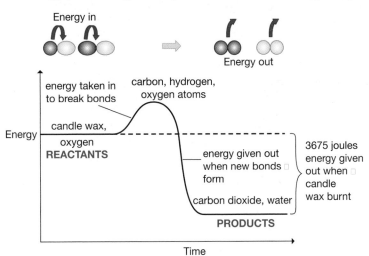

Think about

How would this diagram be different for an endothermic reaction?

Using energy changes

Keeping the UK moving

Network Rail owns 35 400 km of rail track across the UK. Their maintenance workers inspect, maintain and replace rail track. New rails have to be joined together. This may be done by welding. Today, most welding is done by a track-laying machine. In the past, workers used an exothermic reaction to produce molten iron. This filled the gap between the iron rails.

P2 | Useful exothermic reactions

The thermite process

Molten iron for joining rail track was produced using the **thermite process**. A mixture of iron oxide and aluminium powder is piled at the join, using a mould. A fuse of magnesium ribbon is ignited. This starts the reaction:

iron(III) oxide + aluminium $\rightarrow$ aluminium oxide + iron

$$Fe_2O_3 \quad + \quad 2Al \quad \rightarrow \quad Al_2O_3 \quad + 2Fe$$

The reaction is very exothermic. Molten iron at 3000 °C is produced. It welds the ends of the rail track together.

The thermite process is highly exothermic

Know more

The magnesium fuse reacts exothermically with oxygen. The heat given out starts the reaction between iron oxide and aluminium.

🔍 ...thermite reaction

Poly(ethene) has many uses. It is produced in an exothermic reaction

P2

Fireworks

Fireworks contain a mixture of chemicals. Some chemicals provide oxygen. Others are fuel. When the firework is ignited, exothermic reactions take place. Energy is given out, mostly as heat, but some as light. We enjoy the light show.

Making polythene

Poly(ethene) is the correct name for polythene. It is a very useful plastic. It is used to make many things, from plastic bags to bulletproof vests and the coating on skis and sailboards.

Poly(ethene) consists of very long molecules. It is made from ethene gas. The small ethene molecules join up to make the long molecules in a reaction that is very exothermic. The problem is that this heat can speed up the reaction and make it difficult to control. The heat given out has to be conducted away.

M2

An essential endothermic reaction

Plants use **photosynthesis** to make sugar from carbon dioxide and water. The reaction only happens in sunlight. The light provides the energy for the reaction. It is endothermic.

Photosynthesis is an essential endothermic reaction that provides our food

Research

Can you find other endothermic reactions that absorb light energy? Hint: think about photographic film.

Aluminium has many uses

D2

A profitable endothermic reaction

Aluminium is a metal with many uses, from drink cans to aircraft. We extract aluminium from mined **bauxite**, which is aluminium oxide. A lot of energy is needed to break the aluminium–oxygen bonds. They are strong bonds. Metal workers melt the aluminium oxide, then pass electricity through. The reaction is endothermic. It is called **electrolysis.** Electrical energy is used.

🔍 ...ethene

Organic compounds

Organic produce

Organic food is grown without artificial fertilisers. Organic farmers use organic methods such as spreading farmyard manure as fertiliser. Organic clothes are made from fibres like cotton grown organically. 'Organic' has a different meaning in chemistry. Organic compounds are all those that contain carbon. Organic chemistry is the study of these compounds.

P3 Carbon and crude oil

Carbon atoms have four electrons in their outer shell. They can make four covalent bonds, as they do in diamond. But carbon atoms can also join up with each other in long chains. These can be thousands of carbon atoms long. The chains may join up with other atoms to make very big molecules. Some of these molecules are the basis of life. They are **organic compounds**.

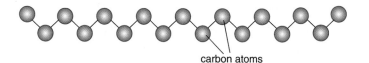

carbon atoms

Long chain of carbon atoms

We have many uses for organic compounds. Plastics, dyes, medicines and foods are a few. We get some of these from plants and animals. We make many more organic compounds from **crude oil**.

Crude oil is a mixture of **hydrocarbons**. Hydrocarbons are molecules that contain only carbon and hydrogen. They are organic compounds. Before we can use these organic compounds, we have to separate them out from the mixture. This is done in an **oil refinery**.

Know more

Biodegradeable plastics are made from corn plants.

...fractional distillation

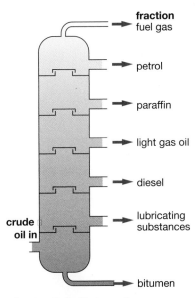

A fractional distillation column

P3

The hydrocarbons in crude oil have different lengths of carbon chain. Some are short and lighter, others longer and heavier. A **fractional distillation column** sorts the hydrocarbons by the size of the molecules.

The lighter molecules rise to the top of the column. The heavier molecules stay at the bottom. Substances with different sized molecules are collected at different places. The different substances collected are called **fractions**.

M3

The petrochemical industry

Some fractions – for example petrol and diesel – are used as they are. Other fractions are used in the **petrochemical industry**. They become the reactants for many different chemical reactions. There may be many steps. The eventual products are the materials we rely on like detergents, medicines, dyes, cosmetics, food additives – the list is endless.

One of the many products of the petrochemical industry

Research

Crude oil is non-renewable. What are scientists doing to find alternatives to oil products?

Think about

How would your life be different if there were no more coal or oil?

Discuss

What environmental issues should petrochemical workers consider?

D3

Bad points

Oil and coal formed millions of years ago. Millions of tonnes of carbon were locked up in rocks. Here they could not cause global warming. Now we take oil and coal out of the ground. We use it as fuel and to make materials. The carbon usually ends up as carbon dioxide in the air. This causes global warming.

...petrochemical industry

Alkanes and alkenes

Your assessment criteria:

P3 Carry out experiments to identify organic compounds

Practical

M3 Describe the uses of organic compounds in our society

D3 Explain the benefits and disadvantages of using organic compounds in our society

High-performance cars

Racing cars and some sports cars have high-performance engines. They need a fuel to match. Petrol fuels contain a blend of hydrocarbons called alkanes. Different blends have different 'octane ratings'. For example, 95-octane Premium Unleaded has an octane rating of 95. It is suitable for most family cars. High-performance cars need higher octane ratings. Using the right fuel means the fuel will burn in a controlled way.

High-performance cars need high-octane fuel

P3

Alkanes

The hydrocarbons in crude oil are **alkanes**. Alkanes are a family of organic compounds. The smallest alkane molecule has one carbon atom bonded to four hydrogen atoms. It is methane. CH_4 is its **molecular formula**. The **displayed formula** shows all the atoms and bonds.

Alkane	Molecular formula	Displayed formula
methane	CH_4	H | H—C—H | H
ethane	C_2H_6	H　H |　| H—C—C—H |　| H　H
propane	C_3H_8	H　H　H |　|　| H—C—C—C—H |　|　| H　H　H
butane	C_4H_{10}	H　H　H　H |　|　|　| H—C—C—C—C—H |　|　|　| H　H　H　H
pentane	C_5H_{12}	H　H　H　H　H |　|　|　|　| H—C—C—C—C—C—H |　|　|　|　| H　H　H　H　H
octane	C_8H_{18}	H　H　H　H　H　H　H　H |　|　|　|　|　|　|　| H—C—C—C—C—C—C—C—C—H |　|　|　|　|　|　|　| H　H　H　H　H　H　H　H

Know more

Organic compound names follow patterns. Meth- means one; eth- means two; prop- means three; but- means four; pent- means five; oct- means eight. It's the number of carbon atoms. The ending ane tells you it's an alkane.

...methane

Know more

Propene is C_3H_6, butene is C_4H_8, pentene is C_5H_{10}. They all have one double covalent bond between two of the carbon atoms. The rest are single covalent bonds.

Bottled gas used for gas barbecues is butane

Displayed formulae are two-dimensional. They do not show the real shape of the molecule. Models like 'molymods' (shown below) show the molecule in three dimensions.

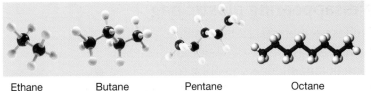

| Ethane | Butane | Pentane | Octane |

Alkenes

Alkenes are another family of hydrocarbons. The first alkene has two carbon atoms joined with a double covalent bond, and four hydrogen atoms. It is ethene. Its molecular formula is C_2H_4. Its displayed formula is:

$$\underset{H}{\overset{H}{>}}C=C\underset{H}{\overset{H}{<}}$$

Ethene

Think about

How can you work out the formula of an alkane?
Crude oil has alkanes with up to 70 carbon atoms. What is the formula of an alkane with 20 carbon atoms?

Using alkanes and alkenes

Alkanes are not very reactive. But they do make good fuels. Methane is the main gas in natural gas. Octane is one of the alkanes in petrol. We use propane and butane in bottled gas including Calor gas ('Camping Gaz') and in **LPG fuel**. Some cars use LPG fuel instead of petrol.

The alkene ethene is used in the petrochemical industry to make plastics such as poly(ethene) and other synthetic materials.

Both petrol and LPG fuel are sold here

The government is backing LPG

1300 filling stations around the UK sell LPG fuel. There are several reasons why people have LPG cars:

- running costs are cut by up to 50%
- there is 20% less carbon in the exhaust gases
- the engine is quieter
- we have plenty of LPG – we even export it
- the road tax is lower
- LPG has less government duty than petrol or diesel, so it is cheaper.

...LPG

Plastics

The disappearing plastic bag

At some supermarkets, cashiers hide their plastic bags under the checkout. They encourage shoppers to bring their own bags or buy their 'green' bags. Some shops now charge for plastic bags. Some towns have banned them all together. But billions of them are still given away each year in the UK. They can be recycled, but many end up in landfill. They take 1000 years to decompose.

P3 Making plastics

Plastics have very long molecules. They are made by joining small molecules end to end. The small molecules are called **monomers**. The long plastic molecules are called **polymers**. Different monomers make different plastics.

Plastic bags are made from poly(ethene). Poly(ethene), also called polythene, is made from ethene. Ethene is the monomer. High pressure and a catalyst are used to make the ethene molecules join up into the polymer molecules. This is called **polymerisation**.

Ethene is:

$$\begin{array}{c} H \\ \end{array} \begin{array}{c} H \\ \end{array}$$
C=C

Poly(ethene) is:

$$\left[\begin{array}{cccccc} H & H & H & H & H & H \\ | & | & | & | & | & | \\ C & C & C & C & C & C \\ | & | & | & | & | & | \\ H & H & H & H & H & H \end{array} \right]_n$$

The *n* means the formula is repeated many times.

The double bond in ethene breaks open and joins with another ethene molecule. Poly(ethene) molecules can be 10 000 carbon atoms long. Poly(ethene) is an **addition polymer**.

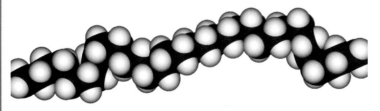

Model of a poly(ethene) molecule

Know more

Poly(ethene) was discovered by accident in 1933 when an experiment went wrong. Several explosions later, scientists realised they had made poly(ethene).

🔍 ...uses of poly(ethene)

Know more

F is the group 7 (halogen) element fluorine. See Unit 1.

P3

Teflon™ is a plastic made from the monomer tetrafluoroethene.

Tetrafluoroethene is:

$$\begin{array}{c} F \\ \diagdown \\ F \diagup \end{array} C = C \begin{array}{c} \diagup F \\ \\ \diagdown F \end{array}$$

When it polymerises, the polymer poly(tetrafluoroethene) or PTFE for short, trade name 'Teflon', is made.

Teflon is:

$$\left[\begin{array}{cccccc} F & F & F & F & F & F \\ | & | & | & | & | & | \\ C - C - C - C - C - C \\ | & | & | & | & | & | \\ F & F & F & F & F & F \end{array} \right]_n$$

The n means the formula is repeated many times.

A slippery plastic

M3

The carbon–fluorine bonds in Teflon are very strong bonds. It makes Teflon very unreactive and slippery. In fact, Teflon has the lowest friction of any material we know. It is used to coat space suits, because it protects astronauts from scrapes, it will not react with the environment and it is heat resistant. Teflon is also the non-stick coating on saucepans and frying pans.

Discuss

Teflon is also used to insulate electrical cables in spacecraft. What property must it have to make it suitable for this?

Research

Teflon has many other uses. Research these.

Designer polymers

D3

Polymers can be built to order. Rather like building with Lego bricks, we can pick monomers to give the right properties. Plastics can be designed to conduct electricity, dissolve in water, change colour or shape. Many companies will design a plastic to do a job.

For example, some medicines are best released slowly in the body. Scientists have designed plastic coatings for tablets. After the tablet is swallowed the plastic releases the medicine slowly.

Slow-release tablets

PVC and PVCu

Your assessment criteria:

P3 Carry out experiments to identify organic compounds

Practical

M3 Describe the uses of organic compounds in our society

D3 Explain the benefits and disadvantages of using organic compounds in our society

A fashion statement

The fashion industry uses PVC fabric as a cheaper alternative to leather. It is made into clothes, bags, belts and shoes. PVC is a plastic. It can be bonded to cotton material or used on its own.

P3

Making PVC

PVC stands for **p**oly**v**inyl**c**hloride. It is the old name for poly(chloroethene). The monomer used to make PVC is chloroethene.

It is an organic compound made from crude oil. When it polymerises, the double bond breaks open and the monomers join up in a long chain.

chloroethene

PVC

The PVC made by polymerisation is hard and rigid. Technicians use **nanochemistry** to make the polymer flexible. Small molecules called **plasticisers** are added. These act as molecular ball bearings. They fit between the polymer chains and allow the long polymer chains to slide over each other. The PVC is now flexible and can be used as a leather substitute in the fashion industry.

Cling film is also made from PVC. Plasticisers make the PVC flexible

We have many uses for PVC. Plastic window frames, guttering and drainpipes are just a few. The PVC used to make window frames needs to be rigid. It does not need plasticisers. This PVC is known as **PVCu** or unplasticised PVC.

> **Know more**
>
> *uPVC, PVCu and PVC-U are the same thing. It was originally uPVC in the UK. The U was moved to the end of the name to fit in with the rest of Europe.*

> **Know more**
>
> *Cl is the group 7 (halogen) element chlorine.*

> **Know more**
>
> *PVCu is rigid because the long polymer chains attract each other. This **intermolecular force** holds the chains in place and the plastic is rigid.*

...intermolecular forces

Why PVCu?

M3

Builders choose PVCu windows, guttering and drainpipes because:

- it does not rot or rust
- it resists weathering – it is not affected by rain, sun or frost
- it is tough
- it keeps its shape at normal temperatures
- it can be reshaped at high temperatures – this means it can be recycled.

85% of all window frames are now made from PVCu

Think about

What other materials can be used to make window frames? Which is best?

PVC problems

D3

Environmental groups like Greenpeace have many concerns about PVC.

- The raw materials used to make PVC come from crude oil. Crude oil is a finite resource.
- Making PVC produces harmful chemicals called **dioxins**.
- The plasticisers in PVC may leak out, and could pose health risks . Many children's toys are made from PVC.
- PVC is difficult to recycle.
- Burning PVC makes more dioxins.

Are we exposing children to too much PVC?

Discuss

Car manufacturers often use PVC for the car upholstery. What are the good and bad points about using PVC?

...Greenpeace

Alcohols

Your assessment criteria:

P3 Carry out experiments to identify organic compounds

Practical

M3 Describe the uses of organic compounds in our society

D3 Explain the benefits and disadvantages of using organic compounds in our society

Brewing beer

We have been brewing beer in Europe for 5000 years. Today, it is a skilled job. The percentage of alcohol is carefully monitored. The alcohol in beer is ethanol. It is made by fermenting sugar from barley. Hops may be added to give a bitter flavour.

P3 The alcohol family

Alcohols are a family of organic compounds. **Ethanol** is just one example.

Alcohols contain carbon, hydrogen and oxygen. A molecule of ethanol has two carbon atoms, six hydrogen atoms and one oxygen atom. Its molecular formula is C_2H_5OH.
The displayed formula is:

$$H-\underset{\underset{H}{|}}{\overset{\overset{H}{|}}{C}}-\underset{\underset{H}{|}}{\overset{\overset{H}{|}}{C}}-O-H$$

Ethanol

All alcohols contain an oxygen atom bonded to a hydrogen atom. It is called the **OH group**.

Beer contains between 3 and 5% ethanol, depending on the type. Wine contains between 9 and 14% ethanol.
Spirits like vodka contain up to 40% ethanol.

Know more

The letters 'eth' tell us there are two carbon atoms.
'ol' is the ending used for alcohols.

Know more

There are many alcohols. Methanol is CH_3OH. Propan-1-ol is C_3H_7OH.

🔍 ...fermentation

Making ethanol

The first step in brewing is changing the starch in barley to sugar. **Yeast** is then used to ferment the sugar. Yeast is a single-celled living organism. It uses the sugar as food. It produces ethanol and carbon dioxide. This is called **fermentation**. The carbon dioxide gas bubbles off and the ethanol stays in the mixture.

Fermentation tank in a brewery

Ethanol in cosmetics

Ethanol is a colourless liquid that boils at 78 °C. It has many uses in cosmetics.

- It is used in acne treatments because it is **antibacterial**.

- It is used to dissolve other ingredients.

- It is used in skin toners and aftershave. It evaporates quickly so the skin feels cool.

An important property of alcohols

Alcohols burn in oxygen. They are **flammable**. The reaction is:

$$C_2H_5OH + 3O_2 \rightarrow 2CO_2 + 3H_2O$$

This is an exothermic reaction: heat is given out. Alcohols make good fuels.

Replacing petrol

Ethanol can be used instead of petrol in vehicles. It reduces our dependence on crude oil. The US and Brazil grow lots of corn and sugar cane. They ferment some of it to make ethanol, often called **bioethanol**. The ethanol is mixed with fuel diesel. Most cars in the US can use blends that are 10% ethanol.

M3

Running cars from plants

D3

Bioethanol is a **renewable** source of energy. It is made from plants and we can grow more plants.

Carbon dioxide is used when plants photosynthesise. It is released when the ethanol fuel is burnt. It balances out. Making bioethanol is a sustainable industry.

carbon dioxide, water, sunlight —photosynthesis→ sugar in plants —fermentation→ ethanol —combustion in cars→ carbon dioxide, water, energy

There is a drawback. Huge areas of land are needed to grow plants for bioethanol. We also need this land to grow food.

Discuss

*Ordinary cars in the UK can run on 5% blends of ethanol. The EU has set targets for all fuels to contain some **biofuel**. We make bioethanol from sugar beet in the UK. Should the government encourage greater use of bioethanol fuel?*

...bioethanol

Carboxylic acids

The sweet shop

The confectioner makes sweets of many different flavours. Fruity flavours come from chemicals called esters. Esters occur naturally in fruits and flowers. Synthetic esters can be made in a laboratory. Scientists can make esters that smell and taste like bananas or strawberries, or any other fruit flavour.

P3

Organic acids

The **esters** that give us fruity flavours are all made from an alcohol and an **organic acid**. The organic acid called ethanoic acid is used to make pear and banana flavours.

Ethanoic acid:

- is a **carboxylic acid** (this is the chemical family)
- has the molecular formula CH_3COOH
- has the displayed formula:

Ethanoic acid

All carboxylic acids include the group:

This is the **carboxylic acid group**.

Making esters

The general equation for making an ester is:

alcohol + carboxylic acid ⟶ ester + water

A few drops of sulphuric acid are needed as a catalyst.

> **Know more**
>
> *Ethanoic acid is corrosive.*
>
>

> **Know more**
>
> *Methanoic acid is HCOOH. Ant stings are itchy because they inject you with methanoic acid.*

🔍 ...artificial flavours

P3

The food scientist makes different esters by using different alcohols and acids. They all result in different flavours. The table shows the reactants needed to produce two particular esters.

Alcohol	Acid	Ester	Smell
ethanol	ethanoic acid	ethyl ethanoate	pear
butanol	ethanoic acid	butyl ethanoate	blueberry/banana

Esters are also used in perfumes. They give the fruity or flowery smells. The ester must evaporate easily, that is, be **volatile**. Then enough particles reach your nose for you to detect the smell. Expensive perfumes may have natural esters such as in rose oil or sandalwood. Synthetic esters are used in cheaper perfumes, including those used in bath salts, room sprays, perfumed fabric fresheners and polishes.

Know more

Ester names all end in -oate.

Production of esters-based perfumes is big business

M3

Vinegar

We have been using a dilute solution of ethanoic acid for centuries. It is vinegar. Vinegar is made when certain bacteria mix with wine, beer or cider. The bacteria make the ethanol react with more to oxygen and this makes ethanoic acid. It happens naturally when wine or beer goes off.

We add vinegar to our food. We use it in cooking. We use it to preserve food like pickles. The old name for ethanoic acid is acetic acid. You will still find this name used on the ingredients lists of some foods.

Onions pickled in vinegar

Think about

Ethanoic acid behaves like other acids. How would you expect it to react with:

- universal indicator solution
- an alkali
- a metal
- a carbonate?

D3

The good and the bad

If the food label says **flavouring**, you may be eating esters made in a laboratory. If the food label says **natural flavouring**, the flavouring was made by nature. It's the same chemical, just made in different ways. Artificial flavourings do not need E-numbers. Many doctors are worried that artificial flavourings cause hyperactivity and allergies in children.

Nutritional food label

Discuss

Artificial flavours do not have to be listed separately on ingredients lists. Sweets and soft drinks can contain many flavours. Should they be listed? Should artificial flavourings have E-numbers?

...vinegar

Nanostructures

Your assessment criteria:

P4 Identify applications of specialised materials

M4 Describe the production of specialised materials

D4 Explain the implications of nanochemistry

Green rocket fuel

One big problem with space travel is getting off the ground. Space technicians are developing a new rocket fuel. It is a mixture of 'nanoscale' aluminium powder and ice. It doesn't work with ordinary aluminium powder. The particles are too big. Nanoscale aluminium has very small particles. They react explosively with water. The products are less polluting than ordinary rocket fuel.

P4

How small?

We use kilometres, metres, centimetres and millimetres to measure everyday sizes. **Nanometres** are used to measure very small things. One nanometre (1 nm) is one billionth of a metre. The smallest bacteria we know are 200 nm long. A gold atom is 0.28 nm across.

Sizewise, a nanometre to a metre is like a marble to the Earth.

Size	How many metres?
kilometre (km)	1000
metre (m)	1
centimetre (cm)	0.01
millimetre (mm)	0.001
nanometre (nm)	0.000 000 001

Chemistry usually works on a large scale. But nanochemistry is about making individual atoms and molecules work for us. It is chemistry on the **nanoscale**. Nanochemistry is different to ordinary chemistry. This is because single atoms and molecules have different properties to bulk substances.

Know more

We could measure very small things in metres.
But the numbers get difficult to handle. A gold atom is 0.000 000 000 28 m across.

Know more

Nanochemistry couldn't be studied until we had a microscope powerful enough to see single atoms.

...carbon nanotubes

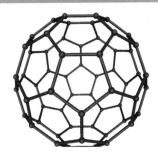

The buckyball 'buckminsterfullerene'

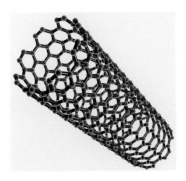

A carbon nanotube

Useful nanostructures

P4

Shape is important in nanochemistry. We used to think carbon could be either diamond or graphite. Then **fullerenes** were discovered. These are different structures of carbon atoms. Two fullerenes used in nanochemistry are **buckyballs** and **nanotubes.**

Fullerene	Shape
buckyball	hollow, spherical (ball-like)
carbon nanotube	hollow, cylindrical (tube-like)

The first buckyball discovered was buckminsterfullerene, C_{60}. It has 60 carbon atoms covalently bonded in a football pattern. It is 0.7 nm across. Other buckyballs have been discovered.

Carbon nanotubes have many formulae. Their diameters are measured in nanometres, but they can be up to several millimetres long. Carbon nanotubes are very strong. They are good conductors of electricity and heat.

Discuss

We all get unexpected experimental results. Should we try to explain them or decide something went wrong? What would have happened if the scientists at Sussex had binned their results?

Discovering buckyballs

M4

Scientists at Sussex University were looking at materials in stars. They found carbon molecules with a formula C_{60}. At first they thought their apparatus wasn't working properly. But they kept getting the same result. They later found C_{60} existed on Earth, in candle soot. They explained the shape of the structure using a football. They received a Nobel Prize for their work in 1996.

Research

Nanotubes have many other possible uses. Find out about some of these.

Think about

Explain the meanings of these words:
nanochemistry   nanoscale
nanostructure   nanometre
nanotubes

The way ahead

D4

Discovering buckyballs and carbon nanotubes was very exciting, but scientists wanted to find uses for them. Some suggested uses were:

- Drugs could be put inside nanotubes. This would enable a cancer drug to be delivered straight to the tumour.

- Carbon nanowires, just a few nanometres wide, could enable the production of nanoscale electronic circuits.

- Buckyballs or nanotubes could act as molecular ball bearings.

This is nanochemistry.

Using nanochemistry

Your assessment criteria:

P4 Identify applications of specialised materials

M4 Describe the production of specialised materials

D4 Explain the implications of nanochemistry

Smelly socks

A mixture of sweat and bacteria can make your socks smell. Anti-odour socks are treated with nanosilver. The tiny silver particles penetrate the fibres and stop the bacteria forming. Your feet may sweat, but your socks will stay fresh. **Nanosilver** is also being used in shirts and sportswear to keep them odour free.

P4 Everyday nanoparticles

Many cosmetics and other everyday items now contain **nanoparticles** (nanoscale particles).

Sun creams

Ultraviolet (UV) light burns your skin. Most sun creams contain titanium(IV) oxide and zinc oxide. These reflect the harmful UV light. They also make a thick, white, greasy layer on your skin. But in some sun creams, the titanium(IV) oxide and zinc oxide are ground to nanoparticles. The 'cream' now appears transparent, but still works.

Mascara

There are many different carbon buckyballs. Their colours range from yellow/orange to brown and black. It all depends on the number of carbon atoms in the structure. Buckyballs make good mascara colours. No extra dye is needed. They easily roll on the eye lashes and are very soft to touch.

Buckyball mascara needs no dye

Know more

Nanoparticles of copper are being used in some self-tanning lotions. They penetrate the skin and your tan lasts longer.

...anti-odour socks

P4

Know more

Some high-street fashion retailers are using nanochemistry to make their clothes stain resistant and wrinkle resistant.

Textiles

Nanoparticles can be used to treat fabrics, to make them:

- antibacterial
- stain resistant
- wrinkle resistant
- flame retardant
- waterproof
- resistant to UV light.

In fact, almost any property you need can be created, even colour. Carbon nanotubes can be spun into fibres. Different thicknesses of nanotubes are different colours. The fibres do not need to be dyed. You just choose the right thickness of nanotubes.

Sports equipment

Nanochemistry has improved sports equipment. Golf clubs can have nanoscale metal coatings to make them stronger. Golf balls can also have nanoscale coatings, to make them go straighter. Tennis rackets with nanoparticles of silicon(IV) oxide have more power. Carbon nanotubes make badminton rackets, ice hockey sticks and baseball bats lightweight and very hardwearing.

Research

A battery with silicon nanowires would last ten times longer in your mobile before it needed recharging. Find out more about nanobatteries.

M4

Chips or atoms?

Computers store information on silicon chips. Computer designers are now looking into nanochemistry. Information could be stored using atoms instead of chips. They call them atom computers or **quantum computers**. They would be much smaller than today's computers – perhaps the size of a sugar cube.

Discuss

What advantages would quantum computers have?

Discuss

Ordinary silver has been tested and is safe to use. Nanosilver has different properties. What tests should all nanomaterials have? What regulations do we need?

D4

The concerns

Safety checks on nanomaterials are not yet regulated. Nano particles are small enough to penetrate your skin, and they can move round your body. Health workers are concerned. They do not know what the effects are.

Nanosilver is slowly washed out of clothes. It kills twice as many bacteria as bleach. The water authorities are concerned it could stop bacterial breakdown in sewage treatment.

Bacteria break down sewage into harmless products. Nanoparticles could affect this action

New materials

A futuristic film set

A futuristic film must use state-of-the-art materials in its props. Costumes must be made from the newest materials with special properties.

P4

State-of-the-art materials

Wool, cotton and silk are natural fibres. We makes use of their natural properties when we make materials from them.

We can design synthetic polymers with almost any special properties we want.

Gore-Tex™

Raincoats can be made from nylon. Nylon is lightweight, tough and keeps the rain out. But it also keeps your sweat in. Water vapour from sweat makes you cold and wet inside the raincoat.

Gore-Tex™ material keeps you dry inside and out. It lets your sweat pass through and keeps the rain out. It has all the properties of nylon, but is breathable as well.

Thinsulate™

Thinsulate™ provides warmth without bulk. Its name is made from the words *thin* and *insulate*. Its polymer fibres are thinner than other insulating materials such as wool. They tangle and trap air. Air is a poor conductor of heat, so you do not lose body heat. The gaps between the fibres also let water vapour from sweat out. You keep dry and warm.

Know more

Lycra™ is 80% polyurethane. It is lightweight, stretchy and used to make sportswear. See Unit 1.

Head-hugging and stylish, but very warm

🔍 ...Gore-Tex, Thinsulate

Know more

New materials used in swimsuits mimic shark skin. Polyurethane sections are water repellent and reduce drag.

Kevlar™

P4

Kevlar™ is a very strong polymer. It is five times stronger than steel. It owes its strength to its chemical structure and how it is woven. The polymer is made and spun into ropes or fibres. Each fibre is made of millions of layers, each one molecule thick. The fibres can then be woven into material. It is used in bullet-proof clothing. The fibre layers can absorb the energy from a bullet or a stab from a knife.

Kevlar has many other uses: to strengthen tyres, to make cables for suspension bridges, sailboard sails, canoes and protective clothing.

Think about

What other uses for Gore-Tex can you think of?

How does Gore-Tex™ work?

M4

Gore-Tex material has at least three layers. The middle layer is Teflon. It has small pores – over 1 billion per square centimetre. These are big enough to let water vapour molecules through. Your sweat escapes. Water drops from rain are too big to pass the other way. You keep dry.

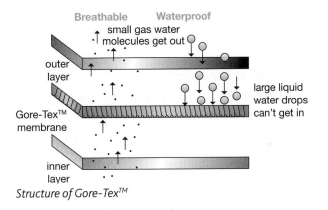

Structure of Gore-Tex™

Think about

How could nanochemistry improve the properties of Lycra sportswear?

Nanotechnology and Gore-Tex™

D4

Nanoparticles are now being used to improve Gore-Tex. When Gore-Tex is treated with carbon nanoparticles, it can be used to make firefighters' jackets. The nanoparticles make it **antistatic**. This means it cannot spark a fire in dangerous conditions. So firefighters can wear it in buildings where gas may be escaping.

Research

Find out how 'cross-linking agents' are used to make new materials.

🔍 ...cross-linking agents

Smart materials

Smart sunglasses

Photochromic lenses in glasses and sunglasses darken as the sun gets brighter. When the sun dims or you go indoors, the lenses become clear. Photochromic lenses are smart materials.

P4

What are smart materials?

Smart materials change with a stimulus. The stimulus must be a change in the environment, like light, temperature, stress or pressure. In **photochromic** lenses, the stimulus is UV light. The change is the amount of visible light that is let through. The change is **reversible**. The lenses can go darker, then lighter again as the light changes.

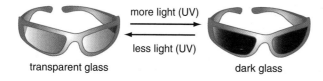

more light (UV)

less light (UV)

transparent glass   dark glass

There are other types of smart materials.

In **thermochromic** materials, the stimulus is heat. The change is their colour. The colour change happens at a definite temperature. The change is reversible. Thermochromic substances can be added to dyes, paint, paper and inks.

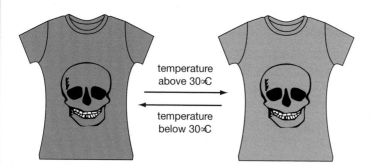

temperature above 30∘C

temperature below 30∘C

Thermochromic dyes in T-shirts change colour at about 30 °C. When you first put the T-shirt on, its temperature is below this. The T-shirt warms up with your body heat and its colour changes when it gets over 30 °C

Know more

Thermochromic dyes in the plastic of babies' bottles can show when the milk is cool enough for the baby to drink.

Thermochromic dyes are used in 'mood' rings. The ring changes colour with the temperature of your finger. It is supposed to show how you are feeling.

🔍 ...transition sunglasses

P4

Piezoelectric materials produce a voltage when they are under stress. This means that squeezing a piezoelectric material and changing its shape makes electricity. The change is reversible: a small voltage across a piezoelectric material changes its shape.

Piezoelectric materials are used in microphones on electric acoustic guitars. They are found under the bridge. They pick up the vibrations from the strings and produce a voltage. The voltage is detected and you hear the music through an amplifier.

Think about

Car windscreens do not let UV light pass though. How useful are photochromic glasses for driving?

Can you suggest other uses for photochromic substances?

Producing photochromic lenses

M4

Lenses in sunglasses are plastic. To make them photochromic, they are dipped in either silver chloride solution or silver bromide solution. The lenses absorb some of the silver compounds. When these silver compounds are exposed to UV light, they change shape. This causes them to absorb some of the visible and UV light falling on them.

UV light is normally only found outside, in sunshine. So photochromic lenses darken outside.

Nanochemistry and smart materials

D4

Nanochemistry is about using atoms as building blocks. We can build a material atom by atom, to do the job we choose. We could put together many more smart materials.

Piezoelectric substances used on the nanoscale could make a **nanomotor**. It could drive a nanoscale vehicle, possibly with buckyball wheels. This could reach places that normal-scale things can't. It could even be sent round your body to treat an illness.

A nanoscale vehicle

Discuss

If you could use nanochemistry to design a new smart material, what properties would it have?

...nanomotor

To achieve a pass grade, my portfolio of evidence must show that I can:

Assessment Criteria	Description	✓
P1	Carry out experiments to identify compounds with different bonding types	
P2	Carry out experiments to investigate exothermic and endothermic reactions	
P3	Carry out experiments to identify organic compounds	
P4	Identify applications of specialised materials.	

Ethane molecule

To achieve a merit grade, my portfolio of evidence must show that I can:

Assessment Criteria	Description	✓
M1	Describe the properties of chemical substances with different types of bonds	
M2	Explain the temperature changes that occur during exothermic and endothermic reactions	
M3	Describe the uses of organic compounds in our society	
M4	Describe the production of specialised materials.	

To achieve a distinction grade, my portfolio of evidence must show that I can:

Assessment Criteria	Description	✓
D1	Explain why chemical substances with different bonds have different properties	
D2	Explain the energy changes that take place during exothermic and endothermic reactions	
D3	Explain the benefits and disadvantages of using organic compounds in our society	
D4	Explain the implications of nanochemistry.	

LO

Be able to investigate motion

- The fastest car in the world is Thrust SCC; it travelled faster than the speed of sound along salt flats in Nevada, in 1997

- The Space Shuttle must fly at 17,500 miles per hour to remain in orbit at about 200 miles above the Earth's surface

- If a car crashes at 45 miles per hour (70 km/h), 1 in 10 drivers die; if the car crashes at 55 miles per hour (90 km/h), 8 in 10 drivers die

LO

Be able to investigate forces

- Satellites orbit Earth above the atmosphere so there are no drag forces; they do not need to burn fuel to stay moving in their orbit

- Formula 1 cars decelerate 60 times quicker than trains, and stop in one-sixtieth of the distance if they travel at the same speed at the start

Be able to investigate light and sound waves

- The mosquito sound is an irritating high-pitched sound that can only be heard by young people; some use it as a silent ring tone on their mobiles; some shopkeepers play it to deter youngsters from gathering near their shops

- Lasers produce very intense beams of light; their uses include tattoo removal, CD players, barcode scanning, pointers, medical operations, very precise industrial cutting

- The use of lasers must be controlled as they can burn human tissue

Be able to investigate electricity

- The first human life saved using a heart defibrillator was in 1947; modern portable defibrillators can check whether the patient needs an electric shock, as well as deciding what type of shock is needed and give instructions to the user

- Many electrical sensors are used in intensive care wards; they can check a patient's oxygen levels, internal fluid pressure, heart activity as well as carrying out tasks like breathing for the patient and keeping them at a constant temperature

Measuring motion

Speeding

If a car is travelling fast, the driver has less time to react if a person steps into the road or a car pulls out. Speed cameras detect motorists who are travelling over the speed limit. People caught speeding are fined and may receive penalty points on their license.

P1 Measuring motion

The **speed** of an object measures how fast it travels. Trains can travel at about 190 km per hour; people walk at about 6 km per hour. Speed is calculated using: speed = distance ÷ time

Speed cameras take two photographs, half a second apart. The photographs show lines on the road and the car's change in position in that half second.

We can also measure average speed over a whole journey:
average speed = total distance travelled ÷ total time taken

If a train takes 3 hours to travel 300 km between two cities, its average speed is 100 km/h. Although it has a top speed of nearly 200 km/h, the train spends some of the journey time stopped at stations.

Displacement is the distance moved in a certain direction. If the train travels north for the whole journey, its displacement is 300 km north. If it zig-zags between the two cities so that the total distance covered is 450 km, the displacement is still 300 km north.

Velocity measures how fast something travels in a certain direction. Velocity is calculated using: velocity = displacement ÷ time
The train's average velocity is 100 km/h north.

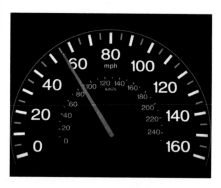

A car's speedometer shows the speed of the car at a particular point in time; the speed changes during a journey

Know more

In the UK, road signs show distances in miles. A speed limit of 30 miles per hour is the same as 50 km per hour (50 km/h).

M1 Motion graphs

A **distance–time graph** is a plot of the distance travelled against the time taken for a journey. The slope of the graph tells us how fast the moving object travels at any point.

- A shallow slope shows slow speed.

- A steep slope shows fast speed.

- A flat line shows the object is not moving.

- If the object moves backwards, the line slopes downwards.

...distance–time graph

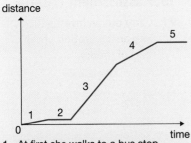

distance

1   At first she walks to a bus stop.
2   Then she waits for a bus.
3   The bus travels quickly at first.
4   The bus slows down.
5   The bus stops.

The distance–time graph on the left shows a girl's journey to school. Each section of the graph has a different slope because she travels at a different speed at different times.

A **velocity–time graph** plots the velocity of a moving object at all times during a journey. Its slope shows **acceleration**. Acceleration measures how quickly the velocity (speed) of an object changes. If the speed is in metres per second, m/s, then acceleration is measured in m/s per second, which is written m/s^2.

- A shallow slope shows slow changes in speed (small acceleration).
- A steep slope shows fast changes in speed (high acceleration).
- A flat line shows a steady speed – no acceleration.
- If the object slows down, the line slopes downwards.

The velocity–time graph below shows a cyclist's journey.

Think about

What would a velocity–time graph look like for your journey to school?

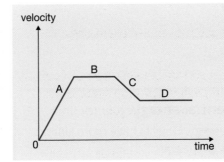

A   The cyclist speeds up.
B   The cyclist moves at the same constant speed.
C   The cyclist slows down.
D   The cyclist moves at a new constant speed.

Think about

In factories, it is necessary to monitor the speed of the production line. How could this be done?

Discuss

*Would car drivers drive more carefully if cars had tachographs? Would drivers drive more safely if all cars had **sat-navs**, which clearly display your speed compared with the speed limit on the road?*

Recording data

Buses, lorries and planes contain equipment to log speeds during all parts of their journeys. This can help detect the cause of any accident, but won't prevent it. Some manufacturers of vehicles fit automatic speed controls and other equipment to detect risky driving before an accident occurs.

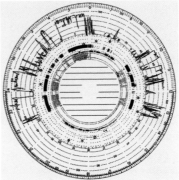

A tachograph logs the motion of the lorry at all times

...velocity–time graph

Kinetic and potential energy

Roller-coaster ride

Many people love the thrill of a roller-coaster. The ride is exciting because you accelerate when you shoot down the track. Potential energy at the top of the track changes to kinetic energy.

Energy changes make this a thrilling ride

P1 | Kinetic energy and potential energy

Moving objects have kinetic energy. Objects have more kinetic energy if they move faster, or are heavier.

Objects that have been lifted up gain energy and store it as potential energy. When an object falls, it changes stored potential energy into kinetic energy. As it falls further it changes more potential energy into kinetic energy and moves faster.

On the roller-coaster ride, energy is conserved – the potential energy changes to kinetic energy at the bottom of each trough, and back to potential energy at each peak.

A child on a swing moves fastest at the lowest point. Then the swing and child have most kinetic energy and least potential energy. When the swing reaches the highest point, it stops moving very briefly before falling again. At this point, all the energy is potential energy. As the child swings, the energy keeps changing between kinetic energy and potential energy.

Sometimes we talk about '**elastic potential energy**'. This is energy gained and stored when an elastic object, such as a balloon, a spring or an elastic band, is stretched. This stored energy can be used to make things move.

Potential energy changes to kinetic energy and back again

Know more

*After a while the swing stops because of energy losses caused by **friction** where moving parts rub, and by **drag** (the continual pushing against the air).*

🔍 ...kinetic energy

M1

Think about

What energy transfers have taken place for a person on a fairground ride so that they have kinetic or potential energy?

Calculating kinetic and potential energy

A moving object has kinetic energy which depends on the mass of the object and its velocity. Kinetic energy (KE) is measured in **joules** and is calculated using:

$$KE = \tfrac{1}{2} \text{ mass} \times \text{velocity}^2$$

with mass in kg and velocity in m/s.

The kinetic energy of Jack (mass 60 kg) walking at a speed of 2 m/s is $\tfrac{1}{2} \times 60 \times 2^2 = 120$ joules.

Objects gain potential energy if they are lifted up. Potential energy (PE) due to height, again in joules, is calculated using:

$$PE = \text{mass} \times \text{acceleration due to gravity} \times \text{change in height}$$

with mass in kg and height in m.

On Earth, acceleration due to gravity is about 10 N/kg.

If Jack climbs up steps so that he is 6 m higher, the potential energy he gains is $60 \times 10 \times 6 = 3600$ joules. The energy needed for this comes from the chemical energy in the food he has eaten.

D1

Making energy changes efficient

When energy changes take place there are always some energy losses, or 'wasted energy'. It is not possible to get rid of all forms of wasted energy, but it is possible to reduce the losses. Moving parts cause friction, wasting energy in the form of thermal energy. Efficient energy transfers reduce this effect, by smoothing or lubricating moving parts. Manufacturers try to design machines to make them as efficient as possible.

Scientists study images like this one of the space shuttle glowing during re-entry to change its design and improve efficiency

Research

Find out some design features that help improve the efficiency of energy changes from kinetic energy to potential energy in leisure equipment, such as trampolines and slides.

 ...potential energy

Forces and motion

Your assessment criteria:

P1 Carry out an investigation into an application of the uses of motion

Practical

M1 Analyse the results of the investigation into the uses of motion

D1 Evaluate the investigation into the uses of motion in our world, suggesting improvements to the real-life application

Stop!

Tyres must be in good condition for a car to stop safely and quickly. A good tread on a tyre helps it grip the road effectively. It reduces the stopping distance if the driver has to brake in an emergency. The fine for a faulty car tyre can be up to £2500.

The tyre on the left is dangerous

P1

Stopping distance

The distance travelled when a car stops is made up from:

- the distance travelled while the driver reacts before braking – the **thinking distance**, and

- the distance travelled while the car is braking – the **braking distance**.

total stopping distance = thinking distance + braking distance

Know more

*The time it takes for the driver to step on the brake after seeing an obstacle is his or her **reaction time**.*

M1

Stopping safely

Cars travelling faster have a longer stopping distance.

thinking distance = reaction time × speed

It is longer:

- if the car is travelling faster – a greater distance is covered during the reaction time

- if a driver is tired, drunk, distracted, or a driver doesn't know the road well – the reaction time is longer.

The braking distance is longer:

- if the car is travelling faster because the car has more kinetic energy to be changed into other forms by the braking forces

- if a car has faulty brakes or worn tyres, or if the roads are slippery, because the braking forces are not as effective.

If a car's speed doubles, braking distance more than doubles, and the danger to pedestrians if they are hit greatly increases. This is because the kinetic energy increases fourfold – it depends on the square of the speed.

Braking distances are longer on slippery roads

...stopping distance

The train's nose is smooth and pointed to reduce drag forces

Moving through fluids

P1

A **fluid** is any substance that flows. Both water and air are fluids. A fluid slows down an object moving in it, because particles of the fluid must be pushed out of the way, and the object rubs against the fluid particles.

This causes a resistive force – generally called drag in any liquid or gas, or **air resistance** in air. The size of the force depends on:

- the speed of the moving object – drag increases with speed
- the type of fluid – liquids cause more drag than gases
- the shape of the object – there is less drag if the object is smooth and has a small cross-sectional area.

Streamlining

M1

Many objects designed to move have shapes that reduce drag forces. Animals like fish and dolphins have a smooth shape with a small cross-sectional area. Designers of submarines and boats have copied these ideas.

Sports cars and high-speed trains travel through air, but at high speeds the drag forces can be considerable. Their fuel consumption increases at high speeds to overcome the drag forces. Their front ends are made wedge-shaped to reduce drag. The benefits are that the vehicle can travel faster and less fuel is needed.

> **Think about**
>
> Think of four or five objects that move fast in water. Identify features they all have that reduce drag.

Improving design

D1

Reducing drag isn't always the most important design aspect. City buses don't travel fast, but must let passengers enter and leave easily. Oil tankers must be able to carry large amounts of oil, and to load and unload easily, but don't need to travel fast. Different jobs mean different priorities for moving objects.

Huge amounts of time and money have been spent designing this car for top speeds

> **Think about**
>
> Some vehicles are more streamlined than others. List five to ten different vehicles and think about their shape. Explain which design feature is most important for each vehicle.

🔍 ... air resistance ... drag force

Stretching and squashing forces

Bungee jumping

Jumping from a tall crane or bridge held by an elastic rope is fun for some people. As you fall, potential energy changes to kinetic energy. The thrill of the jump comes because you accelerate. The elastic rope stretches as you fall, absorbing some energy so you bounce back up.

Your assessment criteria:

P2 Carry out an investigation into an application of the uses of force

Practical

M2 Analyse the results of the investigation into the uses of force

D2 Evaluate the investigation into the uses of force in our world, suggesting improvements to the real-life application

P2 Squashing and stretching

Some objects are elastic – when we stretch or squash them they change shape. When the force is removed, they change back to their original shape. Other objects, like plasticine, are not elastic – once their shape is changed it stays changed.

When you sit on a car seat, you squash the springs inside the seat. We say your weight is a **compressive force** because it compresses, or squashes, the springs. When you stretch an elastic band, we say it is in tension. The forces stretching the elastic band are **tensile forces.**

Sometimes a force is compressive and tensile. If you load a beam, the middle of the beam bends downwards. The underside of the beam is stretched and in tension. The top of the beam is being compressed.

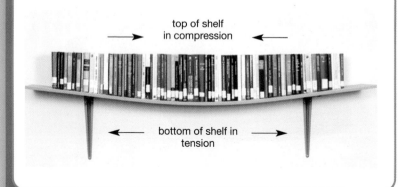

top of shelf in compression

bottom of shelf in tension

Know more

Arch bridges are designed so the weight of traffic squashes each stone onto its neighbour. Stone is very strong when it is compressed.

Know more

Suspension bridges use stretched steel cables to hold the bridge up. Steel is strongest when it is in tension.

...compressive forces, tensile forces

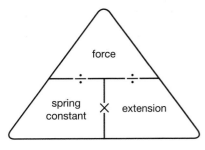

Use this triangle for calculations to do with Hooke's Law. Cover up what you need to find out, and the rest of the triangle shows you how to calculate it

Think about

What is different about spring scales that measure very small items accurately and ones that measure very heavy items accurately?

M2

Hooke's Law

When a force is applied to an elastic object, it changes shape. If objects are stretched along their length, they extend. The amount of extension depends on the force applied. For a spring, Hooke's Law states that:

force = spring constant × extension

The **spring constant** depends on the spring itself. Some springs are much harder to stretch than others; these have a larger spring constant.

The 'extension' means the amount that the spring stretches. If a spring is 10 cm long and extends to 12 cm when stretched, its extension is 2 cm.

This idea can be used to weigh objects. Known weights are hung from the spring which extends. A scale is drawn using this information. This process is called **calibration**. When an unknown mass is hung from the spring, the scale shows its mass.

D2

Good design

When springs are extended too far, they deform and do not go back to their original size. If this happens, then scales would not give a true reading.

If an object that is too heavy crosses a beam or a bridge, the forces may damage it and make it weaker. Engineers need to build in protection.

Every time a new structure or piece of equipment is designed, the forces on it must be thought about, so that it can be used easily without being damaged. This applies to things in the home like door hinges and kitchen scales, as well as larger items like theme park rides and even planes.

This bridge collapsed in 2007, partly due to design flaws

Research

Find out more about building regulations. Why are they necessary and how do they help people to be confident that buildings are safe?

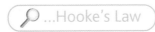
...Hooke's Law

Effects of forces

Your assessment criteria:

P2 Carry out an investigation into an application of the uses of force

Practical

M2 Analyse the results of the investigation into the uses of force

D2 Evaluate the investigation into the uses of force in our world, suggesting improvements to the real-life application

Tug of war

In a tug of war, two teams pull in opposite directions on a rope. If the teams are well matched, both teams pull as hard as each other and the forces are balanced. No team moves much, however hard they pull. In the end, one team gets tired and the other team can pull harder than them, winning the contest.

Matched teams mean the forces are balanced

Know more

*When an object feels a force, it pushes back with an equal force in the opposite direction. If you push on a wall, the wall pushes back on your hand. This is a **reaction force**.*

P2 Balanced forces

Forces can push, pull or twist an object. Usually more than one force acts on an object. We can combine the effects of these forces to give a **resultant force**. Forces are measured in **newtons** (N).

Imagine two friends pushing a heavy object along the floor. Each friend pushes with a force of 40 N.

- Forces acting in the same direction add together. If the friends push in the same direction, they may be able to move the object. Their resultant force is 80 N.

- Forces acting in opposite directions subtract and cancel each other out. If the friends push in opposite directions, the object won't move. The resultant force is zero.

Forces acting on an object that are equal in size but act in opposite directions are **balanced forces**.

Balanced forces mean that:

- an object that isn't moving stays still

- a moving object keeps moving in a straight line at a steady speed.

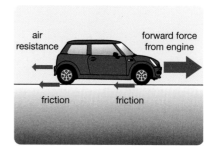

The car travels at a steady speed, because the force from the engine forwards is equal to friction and air resistance backwards

...resultant force

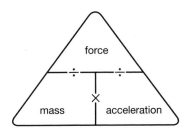

Use this triangle for calculations. Cover up what you need to find out, and the rest of the triangle shows you how to calculate it

Unbalanced forces

M2

Unbalanced forces on an object mean there is a resultant force. This will change the way the object moves.

- An object **accelerates** (speeds up) if the force is in the direction that it's moving.

- An object **decelerates** (slows down) if the force is in the opposite direction to its movement.

- An object changes direction if the force is in a different direction.

A ball thrown upwards slows down because its weight acts downwards – the opposite direction to its movement. Once it starts to fall back down, the ball accelerates because the weight acts in the same direction as its motion.

The amount of acceleration depends on the mass of the object and the size of the force. A large mass needs a larger force to accelerate it than a smaller mass. The same mass accelerates more if a larger force pushes on it. We calculate the acceleration using:

force = mass × acceleration

Force is in N, mass in kg and acceleration in m/s². For example, a runner, mass 60 kg, accelerates by 2 m/s². The force needed is 60 × 2 = 120 N.

Think about

How can a driver control the forces acting on a car so that it reaches different steady speeds?

Changing forces

D2

It can be hard to identify all the forces on an object. But if an object isn't moving, all the forces on it must be balanced.

Some forces change.

- Drag is a force that slows down moving objects. It is larger when the object travels faster. When the drag matches the driving force, the object reaches a top speed, or **terminal velocity**.

- Reaction forces only exist when another force is applied. If you stand on the floor, a reaction force acts upwards in the opposite direction to your weight, so you don't fall through the floor. When you step away, the reaction force disappears.

Research

Find out how drag forces can be controlled in different sports.

...acceleration

Forces on cars

Car crashes

When a car crashes, the driver and passengers feel enormous forces. Crashes at quite slow speeds can still kill and injure people. Modern cars include many safety features to reduce the forces felt by passengers and to make accidents less likely.

Test crashes are done to rate the safety of different cars

P2

Braking

Friction is a force that slows down moving objects. It enables cars to brake. When the brake pedal is pressed, friction between the car's brake pads and the wheel make the wheel turn less quickly, and stop.

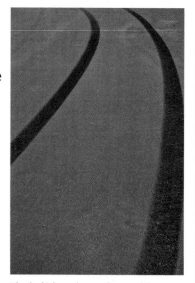

The contact between tyres and the road is very important. When the wheels stop turning, the car also stops if the tyres grip the road well. Otherwise, the wheels stop turning but the car slides across the road. In countries where very cold weather is normal, drivers fit snow chains or winter tyres that grip well in ice or snow.

Black skid marks are from rubber tyres sliding across the road as the vehicle skids. Skids are worse at high speeds

Know more

In icy weather, a mixture of grit and salt is spread on roads. The grit makes the road less slippery and the salt helps the ice to melt.

...friction

M2

Think about

List several occasions when it is likely that a driver will skid.

After a crash, seatbelts cannot be re-used

Braking safely

A braking system called **ABS** helps a driver to keep control of the car if it skids. When the driver brakes, sensors in the brakes check that the wheels are still turning. If they are not, the system repeatedly stops and starts the brakes. This helps the tyres grip the road, allowing the driver to steer.

Keeping passengers safe

When a car crashes, it stops suddenly. But the people inside keep moving until something stops them. A sudden stop involves a very large deceleration, so the person feels a very large force and can be injured. Cars are designed so that these forces are reduced.

- Seatbelts keep people in their seats so they are not thrown through a windscreen or into hard objects in the car.

- Seatbelts are very slightly elastic. When the car stops, the seatbelt stretches slightly as the person is thrown forwards. The person takes a fraction of a second longer to stop, which is enough to reduce the force they feel.

- **Air bags** are normally hidden inside the steering wheel or dashboard. If the car crashes, these pop open, creating a cushion filled with gas. The person is thrown forward but falls onto the cushion which reduces the forces they feel.

- The front and back of a car are designed to crumple a lot if there is a crash. This means the car takes a split second longer to stop and so the forces are reduced. The car's **crumple zones** will be badly damaged but the centre part of the car is less damaged.

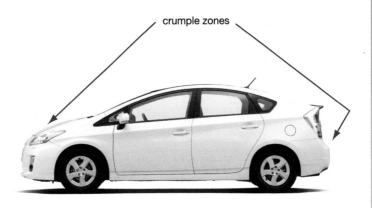

crumple zones

Crumple zones absorb the energy of a crash. Long bonnets absorb more energy as the crumple zone is large

Research

What other safety features are designed in modern cars? Which features do you think are best at reducing injuries and deaths?

🔍 ...car safety features

Parachutes and rockets

Why the *Beagle* didn't land safely

Beagle 2 was a small space probe that was sent to Mars in 2003 to find out more about the planet. The probe was dropped onto the surface of Mars but failed to land safely. Parachutes did not slow it down as much as scientists expected and so it crashed landed, damaging itself in the process.

P2

Slowing down

Drag slows down moving objects. It depends on:

- the object's speed – there is more drag at fast speeds
- the object's surface area – there is more drag if the surface area is large
- the stuff the object moves through – there is more drag if this is thicker, for example, a liquid.

Gravity

Gravity is a force that attracts two objects together. The size of gravity depends on the mass of the objects and their separation. The Earth has such a large mass that it attracts small, nearby objects (like us) to its surface. Our weight is the force of gravity.

It affects the motion of other objects that are further away, like the Moon, but its effect is weaker at a distance.

Other planets and moons all have their own gravity. The Moon's gravity is less than Earth's, because the Moon's mass is less that Earth's.

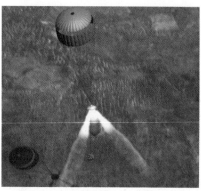

Landing a probe safely on Mars is a challenge for scientists

Know more

The force of gravity on Mars is less than on the Earth because the mass of Mars is less than the Earth's mass.

M2

Parachutes

When any object falls, it feels two main forces: its weight acting downwards and drag acting upwards.

1 When an object starts falling, there is a downwards force so it accelerates. Weight is much bigger than drag.

2 As the speed increases, so does the upward drag. The weight acting down stays the same. The resultant force gets smaller so there is less acceleration.

3 Eventually, there is no resultant force: drag and weight are equal. The object keeps falling, at a steady speed or terminal velocity.

...gravity, terminal velocity

Research

Find out about the world record for the highest sky dive. How was the speed of the dive controlled?

A parachute is used to slow down the speed of a falling object. If it opens while an object falls:

1 Drag increases because the surface area is bigger.

2 The upward drag force is larger than weight. The resultant force is upwards so the object slows down (decelerates).

3 As the speed decreases, the drag get less so the resultant force gets smaller until drag and weight are equal. The object keeps falling but at a slower steady speed (terminal velocity).

In each case the skydiver falls at his terminal velocity, but this is slower when the parachute is open

Huge amounts of fuel are needed so the rocket can break away from the Earth's gravity

Rockets

Beagle 2 was sent off on its mission using a rocket. Rockets leave the Earth so fast that gravity is not strong enough to pull them back. As the rocket moves further away from Earth, the pull of gravity is less so the force needed to move away is less. As the rocket moves closer to Mars, Mars's gravity starts to attract it more than the Earth's gravity.

Research

What effect does the Moon's gravity have on things on Earth?

Landing safely

One theory of why *Beagle 2* fell heavily is that the atmosphere on Mars was not as thick as scientists believed. The parachutes weren't big enough, or didn't open in time, to slow *Beagle* down enough to avoid damage.

...rocket forces

Lenses

Your assessment criteria:

P3 Carry out an investigation into an application of the uses of waves

Practical

M3 Analyse the results of the investigation into the uses of waves

Lighting a fire with a lens

Can you light a fire without matches? Some explorers use a magnifying glass or even the glass from the bottom of a bottle. If the Sun's heat and light is focused through the glass onto dry scraps of paper or wood, they can catch fire. Litter which includes glass can be a fire hazard amongst dry materials.

P3 Refraction and lenses

Light travels incredibly fast in air, but slows down in **transparent** materials like glass. If light enters a glass block along an imaginary **normal** (this means at right angles to the boundary), it keeps going in a straight line. If it enters at another angle, it changes direction. This is called **refraction**.

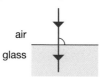

Light entering the glass at an angle of 90°.

The light ray slows down when it enters the more dense glass. It enters the glass along the normal. Its direction does not change.

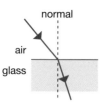

Light enters the glass block at an angle, it slows down in the glass and refracts towards the normal.

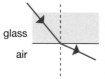

As the light ray leaves the glass block, it speeds up and refracts away from the normal.

Refraction is used by diamond cutters in the jewellery trade to make a diamond appear more impressive

Know more

In empty space, light travels faster than anything else in the Universe – 300 thousand km per second! Even so, it takes light 8 minutes to reach the Earth from the Sun.

...convex lens, concave lens

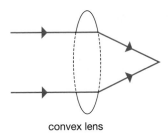

convex lens

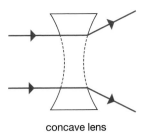

concave lens

Lenses are pieces of glass or plastic, shaped to control the way light travels through them by using refraction.

- A **convex lens** is thicker in the middle than at the edges. It focuses a beam of light to the lens's **focal point**. It is a *converging* lens.

- **Concave lenses** are thinner in the middle than at the edges. They spread light out. It is a *diverging* lens.

M3

Ray diagrams

Light always travels in straight lines called rays. **Ray diagrams** show how light travels from the **object** we are viewing, through the lens to form an **image** that we see. The type of image depends on the position of the object. It may be:

- upside down or upright

- **magnified** (larger) or **diminished** (smaller) than the object

- a **real image** (on the other side of the lens to the object) or a **virtual image** (on the same side of the lens as the object).

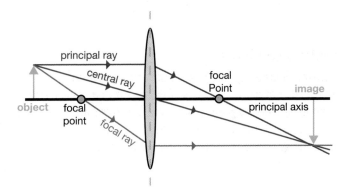

A ray diagram for a convex lens; the image here is upside down, larger than the object, and real

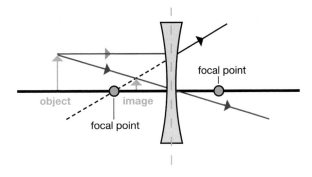

A ray diagram for a concave lens. The image is always upright, smaller than the object, and virtual

Research

Find out the difference between a real image and a virtual image.

Using lenses

Your assessment criteria:

P3 Carry out an investigation into an application of the uses of waves

Practical

M3 Analyse the results of the investigation into the uses of waves

D3 Evaluate the investigation into the uses of waves in our world, suggesting improvements to the real-life application

Seeing clearly

Being able to see clearly can be a matter of life or death. The eyes of animals, birds and fish all contain lenses, but their shapes are different. Some animals, like hawks, are hunters and need to focus clearly on their prey. Other animals, like rabbits and mice, are likely to be eaten. So they need good all-round vision to spot hunters in time. Some animals, like owls, need to see in the dark.

P3

The structure of the human eye

Eyes use convex lenses to focus images onto the back of the eye.

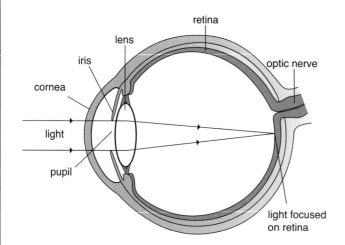

The **cornea** is a transparent covering. It protects the front of the eye as well as working with the fluid in the eyeball to focus light.

The **lens** allows you to focus on near and distant objects by changing its shape. It makes small adjustments so the image on the retina is clear. The lens is thicker when you focus close up. It is thinner when you focus on distant objects.

The **pupil** is the dark spot in the centre of our eyes – it lets light into the eye.

The **iris** surrounds the pupil, controlling its size. This controls how much light goes into the eye. In bright light, the pupil is small.

The **retina** is the covering of light-sensitive cells at the back of the eyeball. These cells detect the image, sending signals along the **optic nerve** to the brain.

Rabbits have eyes on the sides of their heads to help them see all around while grazing

Know more

Our eye sees everything upside down. Your brain makes the changes so you see the world upright.

...structure of the eye

The concave lenses of the spectacles help the eyes to focus on distant objects

Correcting eye problems

M3

Many people do not see clearly because their eyes do not focus light onto the retina.

- Distant objects are blurred if you are **short-sighted**. The image is formed in front of the retina, possibly because the eyeball is slightly too large. To correct short sight, a concave lens is used to spread the rays slightly before they enter the eye.

- Nearby objects are blurred if you are **long-sighted**. The image is formed behind the retina, possibly because the eyeball is slightly too small. To correct long sight, a convex lens is used to bend the rays together slightly before they enter the eye.

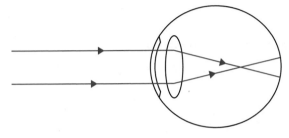

Short-sighted people focus the image of a distant object in front of the retina; a concave lens corrects this

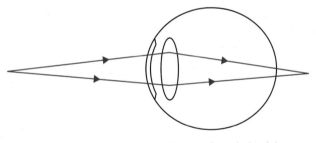

Long-sighted people focus the image of a near object behind the retina; a convex lens corrects this

Laser surgery

D3

Short and long sight can be corrected with surgery. A laser is used to reshape the cornea by burning away cells just under, or at the surface. The shape of the cornea is made more curved (so it focuses more) to correct long sight. The shape is made less curved to correct short sight.

Laser eye surgery has not been in use for long in the UK. This means that long-term effects are not known. As with all surgery, sometimes things go wrong and patients may have permanent eye damage.

Research

Find out more about laser surgery. Is it suitable for all eye problems? Do most doctors think its benefits outweigh its risks?

Reflection

Look behind you!

It's important that drivers know what vehicles behind them are doing. A driver can look behind quickly and safely using mirrors on the windscreen and at the sides of the car. The image is formed from light that has been reflected from the mirror.

P3

Reflection of light

Light reflects (bounces) off almost all surfaces. Light-coloured surfaces reflect more light than dark-coloured ones. We see things because the reflected light enters our eyes. When light is reflected off a flat mirror, the image is:

- the same size as the object

- the same distance from the mirror as the object

- behind the mirror

- a **virtual image** (which means it can be seen, but it can't be projected onto a screen)

- laterally inverted (left becomes right).

When a single ray of light is reflected from a mirror, it obeys the **law of reflection.**

The angle of reflection is the same as the angle of incidence.

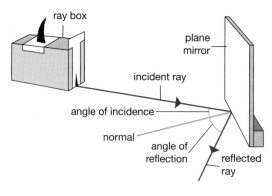

Normal: an imaginary line at right angles to the mirror
Angle of reflection: the angle between the normal and the reflected ray
Angle of incidence: the angle between the normal and the incident ray
Incident ray: the incoming ray
Reflected ray: the ray reflected off the mirror

This shows one way to test the law of reflection

The word 'AMBULANCE' on the front of the vehicle is written backwards, with each letter turned round too. In the mirror of the car in front, the word is seen normally, and the driver knows to get out of the way

Know more

Driving mirrors are convex (they bulge out in the centre). This allows drivers to see more of the car's surroundings. The image is smaller.

...law of reflection

M3

Research

Find out other advantages and disadvantages of reflecting and refracting telescopes.

Telescopes

Telescopes are used to make distant objects appear closer. These objects are usually very dim, so a system of lenses and mirrors is needed to collect the light and concentrate it to give an image bright enough to see.

Refracting (astronomical) telescopes use two convex lenses. The objective lens collects and focuses light from distant objects. The eyepiece lens magnifies the image.

Reflecting (Newtonian) telescopes use a primary concave mirror instead of the objective lens to collect and focus light. They also have a second flat mirror to reflect light to the eyepiece lens. There is less distortion of the image with this set-up.

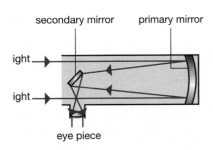

Reflecting telescope

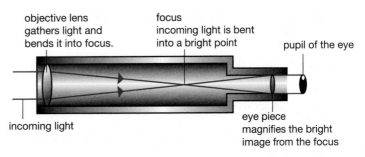

Refracting telescope

D3

The Hubble Space Telescope

The Hubble Space Telescope is a reflecting telescope that is in orbit about 610 km above the Earth. Astronauts carry out its maintenance and any repairs needed. It provides sharper images of objects in space than telescopes based on Earth, partly because the Earth's atmosphere does not interfere with images. It can collect light from very distant, dim objects so it allows us to see further into space than any other telescope has done. It has helped to confirm the age of the universe as 13 to 14 billion years.

This has come at a cost – it has taken billion of dollars to build, service and maintain the telescope. These costs have been shared among several countries, and many people say this is worthwhile as it contributes so much to scientific understanding. Others feel the money could have been spent on schools, hospitals and other community projects.

Research

Weeks after the Hubble Space Telescope was launched, it was discovered that the primary mirror was ground to the wrong shape. The edges were wrong by about two millionths of a metre, in a mirror that was 2.4 m in diameter. Find out why this was a problem, and what was done about it.

...refracting telescope ...Hubble Space Telescope

Internal reflection

Your assessment criteria:

P3 Carry out an investigation into an application of the uses of waves

<u>Practical</u>

M3 Analyse the results of the investigation into the uses of waves

D3 Evaluate the investigation into the uses of waves in our world, suggesting improvements to the real-life application

Optical fibre Christmas trees

One of the latest fashions in Christmas decorations is the optical fibre tree.
The main trunk lights up and light shines out of the ends of the fronds. They look simple, so it's hard to realise how optical fibres have changed our lives, especially in communications and medicine.

P3

Total internal reflection

The inside surface of transparent objects behaves like a mirror when light hits the surface at certain angles. This is called **total internal reflection**. All transparent materials have a **critical angle**.

- If the angle of incidence is smaller than the critical angle, light refracts in the usual way.

- If the angle of incidence is bigger than the critical angle, light reflects off the inside surface.

- Depending on the shape of the object, total internal reflection can occur several times inside it.

Know more

*Diamonds sparkle because their critical angle is very low.
This means that they reflect lots of light inside them. They are cut to make most the most of this effect.*

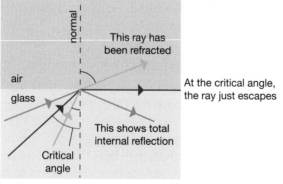

The angle of incidence affects how the light behaves inside a glass block

Optical fibres

Optical fibres are thin fibres made of glass or plastic, about as thick as a single hair. Light entering one end of the fibre can travel many kilometres before it comes out at the far end. This is because the fibre is designed so that the light always hits the inside surface at angles larger than the critical angle. So total internal reflection takes place repeatedly inside the fibre. No light escapes from the sides of the fibre.

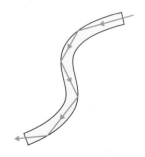

In an optical fibre, total internal reflection takes place many times

 ...total internal reflection, optical fibres

Using optical fibres

M3

Optical fibres enable us to 'see round corners'. They are very useful to doctors for seeing inside the human body. An **endoscope** has thin flexible bundles of optical fibres to shine light into a patient's body. Another bundle of optical fibres sends reflections back so the doctor can view an image of what is happening inside the body. This can be used to diagnose problems, or even to carry out operations using **keyhole surgery** (see Unit 14). Very intense light from a laser can be sent through optical fibres to make cuts or to seal off blood vessels.

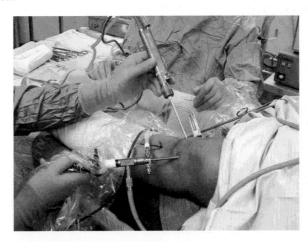

Keyhole surgery allows patients to recover more quickly than after traditional operations

Optical fibres are also used for rapid communications in telephone networks, cable TV and internet connections. Many signals can be carried at the same time down one optical fibre. Cables are thinner and lighter than ever before, but can carry more information.

Optical fibre network cables

Some effects of using optical fibres

D3

Optical fibres have changed medical treatment greatly. However, surgeons need to retrain and practise new techniques to be sure that they can treat patients successfully. Hospital equipment must be updated to keep up with the latest technology.

In the same way, telephone companies have invested millions of pounds in changing many kilometres of copper cables for optical fibres, causing some disruption and extra costs for customers. The benefits are great, but the speed of technological change has had a big impact.

Research

Find out other uses of optical fibres.

...endoscope

Sound and ultrasound

Voice recognition

People with serious eyesight problems, or disabilities with their hands, can use computers with the help of voice recognition software. They speak into a microphone, and the words are changed into electrical signals and saved as a text file.

Your assessment criteria:

P3 Carry out an investigation into an application of the uses of waves

Practical

M3 Analyse the results of the investigation into the uses of waves

D3 Evaluate the investigation into the uses of waves in our world, suggesting improvements to the real-life application

P3

Sound waves

Sound is produced when something vibrates. The vibrations transfer sound energy from one place to another.

- Larger vibrations make the sound louder.

- Faster vibrations make the sound higher pitched.

When something vibrates, it makes nearby particles vibrate. These will bunch up or spread apart, creating regular pressure changes. This causes sound waves to travel away from the vibrating object. Energy is transmitted.

Sound waves must travel through a medium such as a solid, liquid or gas. Sound travels fastest in solids because the particles are closely packed. It does not travel at all in a vacuum (empty space).

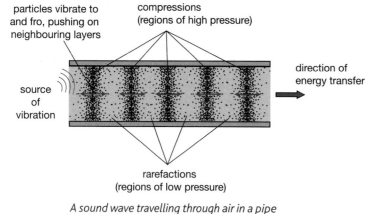

A sound wave travelling through air in a pipe

One vibration per second has a frequency of 1 hertz. Humans hear low-pitched sounds, from about 20 hertz, up to very high-pitched sounds of about 20 000 hertz.

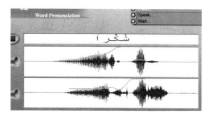

Spoken words represented as an electrical signal

Know more

As we get older, we cannot hear very high-pitched sounds. Some shops try to stop teenagers from gathering nearby by playing very high-pitched sounds that only young people can hear.

...voice recognition software

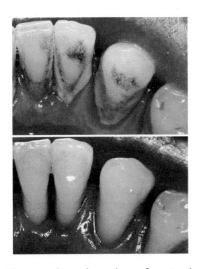

Ultrasound can clean plaque from teeth

P3

Above 20 000 hertz, sounds are called **ultrasound**. Ultrasound is sound that is too high-pitched for humans to hear. Ultrasound waves:

- can be reflected off many materials
- are partly reflected by tissues in the body
- can be directed very precisely
- can break down some substances.

Applications of ultrasound

M3

Ultrasound can be directed as a narrow beam from a transmitter. It sends pulses of ultrasound which are reflected when the beam reaches a boundary between different materials. A receiver measures how long it takes for the reflected ultrasound to return. A pulse travelling further takes longer to reach the receiver.

- In hospitals, ultrasound is used to scan organs like the heart, liver and brain. The pulses reflect in different ways off the different tissues, so a picture of structures inside the body is built up.
- Ultrasound scans in pregnancy can safely look at the baby as it develops inside the mother.
- Ultrasound reflects off cracks in metal, so aeroplane parts, engines and other solid items can be checked.
- In the ocean, ultrasound reflects off the seabed or objects lying beneath the transmitter. The time taken for signals to reach the receiver is measured and used to measure the depth of the seabed or object. The technique can identify fish or whales, and locate the position of sunken ships.

If a beam of ultrasound is directed at some things, it can force them to break apart. It can break down kidney stones, so patients don't need an operation. It can break up plaque which has built up on teeth. It can even dislodge dirt from delicate objects like jewellery.

Research

How do some animals use ultrasound?

Benefits and disadvantages of ultrasound

D3

Ultrasound has been used for many years to examine patients internally without operating. No serious side-effects are known. As ultrasound is becoming more widely used, its safety record is becoming more reliable. However, scans need to be interpreted properly, otherwise doctors may not have the correct information before treating a patient.

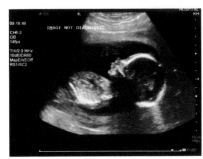

Ultrasound scan of a 20-week-old unborn baby

Electricity and the body

Restarting the heart

The first few minutes after a heart attack are critical. A heart attack stops the heart beating correctly and the beating must be restarted fast. A defibrillator can be used to give the heart an electric shock to restart it. Defibrillators have been installed in places like shopping centres, sports centres, railway stations and airports. This can save lives.

P4

Electric circuits

A current is a flow of electrons round a circuit. A complete circuit of electrical **conductors** is needed for the current to flow. An energy source (a battery or mains electricity supply) provides the electrons with energy. This energy is also called **potential difference** or 'voltage' and is measured in volts. There are two main types of circuit: series and parallel.

A **series circuit** is a single loop of conductors.

- The current is the same in all places in a series circuit because the electrons have only one path they can take.

- The potential difference is shared between the different components in the circuit.

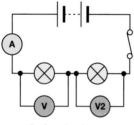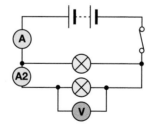

Series circuit   *Parallel circuit*

A **parallel circuit** has more than one loop, forming separate paths that the electrons can go through. Each loop links to the section containing the battery.

- The current is shared between the loops of a parallel circuit because the electrons have a choice of paths.

- The potential difference is the same in each loop. This is because each loop links to the battery.

Know more

Look back at Unit 2 if you need to remind yourself about circuit measurements or circuit symbols.

Know more

Humans can produce about 0.01 to 0.1 volts inside their bodies, but electric eels can produce about 600 volts.

🔍 ...series circuit, parallel circuit

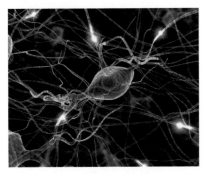

The nervous system uses electricity to pass signals around the body

Circuits and our bodies

P4

Our bodies conduct electricity. Electricity in our bodies:

- makes the heart beat in a regular rhythm
- makes our muscles contract when we want to move
- passes signals to and from our brain using the nervous system.

If a person has an electric shock, it interferes with the body's electrical system. A dangerously large electric current through the body can:

- cause pain and burns
- freeze muscles so the person can't move
- make the heart beat in a dangerous rhythm.

The damage is worse if:

- electricity travels through organs, like the heart or lungs
- the electric shock lasts a long time or is high voltage
- the person (or their clothes) has a low resistance so that a large current flows through them, for example if the skin is wet.

The heart and electricity

M4

The heart should beat in a regular rhythm to push blood around the body. The heart **fibrillates** if it starts to twitch rather than beat, or if it beats too fast. A **defibrillator** gives the heart an electric shock to make it start beating regularly.

Some people's hearts don't beat correctly, so a small defibrillator is placed near the heart to send regular electric shocks and make the heart keep beating steadily.

An **electrocardiogram (ECG)** is a test that records the rhythm and electrical activity of the heart. Patches called electrodes are placed on the person's arms, legs and chest, and connected to a machine that picks up electrical signals. This detects many different heart problems.

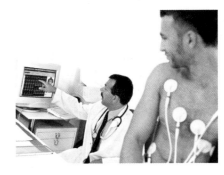

An ECG trace gives useful information about heart problems

Think about

In what other ways can doctors detect heart problems?

Electric sensors

Your assessment criteria:

P4 Carry out an investigation into an application of the uses of electricity

Practical

D4 Evaluate the investigation into the uses of electricity in our world, suggesting improvements to the real-life application

The intelligent house

Intelligent houses automatically shut curtains when it is dark, turn on lights and the TV when people come in, even turn around to follow the Sun or let chosen friends in when you are out. These things are already possible using sensors.

P4

Sensors and indicators

A **sensor** is a device that detects a change in the environment, such as the temperature. In an electric circuit it can cause a change in current, which can make something happen.

Temperature sensors

A circuit using a **thermistor** detects if the temperature is too high or too low. The electrical resistance of a thermistor falls as the temperature increases. This means the current in the circuit gets bigger. So a thermistor is used in circuits which turn things on automatically, such as greenhouse heaters.

A sensor circuit with a thermistor can also turn on an alarm, such as a fire alarm.

Farmers lose money if crops are damaged by frost. A sensor circuit can be designed so that heaters come on when the temperature drops too low

🔎 ...thermistor

Powerful LEDs are used in torches

P4

Light sensors

The resistance of a **light dependent resistor (LDR)** falls as the amount of light falling on it increases. This means the current in the circuit gets bigger. A circuit using a LDR can detect when light levels change. It can be used to turn things on automatically, such as lights when it gets dark.

These lights are controlled by a sensor; they come on when daylight fades

Indicators

A **light emitting diode (LED)** is a very small light that is used as an **indicator** in many circuits. LEDs use a tiny amount of energy and are unlikely to break, so they are used in many items of electrical equipment to indicate whether the circuit is switched on.

D4

Bigger LEDs

More powerful LEDs have recently been developed, which are used in Christmas lights, night-lights and torches. They are reliable and they use a very small amount of energy compared with conventional bulbs.

Because less energy is wasted as heat in an LED, their light is more intense than a bulb with the same power.

LEDs are available in different colours. Their colour is the colour of the light given out, not the colour of a coating on a bulb which can be scratched off in time.

Electric cars

Cars powered by sunshine

Cars can be powered only using energy from the Sun. During the day, arrays of solar cells provide the electricity needed. When it is dark or dull, the car can be driven using batteries that were charged during sunshine. These store the solar energy.

Your assessment criteria:

P4 Carry out an investigation into an application of the uses of electricity

Practical

M4 Analyse the results of the investigation into the uses of electricity

D4 Evaluate the investigation into the uses of electricity in our world, suggesting improvements to the real-life application

P4

Electric cars

Most cars burn fuels – petrol or diesel – to run their engine. Electric cars use an electric battery as a source of energy, instead of fuel. Some electric vehicles have been around for a long time. These include milk floats, buggies for the disabled, and electric trains.

Electric cars have developed recently because batteries have improved. Before this, batteries could not cope with the speed and distances that cars are expected to travel.

Know more

All cars use batteries. The battery in a car is used to start the engine, and to run lights, heating and stereo system.

...electric car

Electric engines are quite small but the batteries take up a lot of the space

Advantages and disadvantages of electric cars

M4

Advantages of electric cars include:

- they do not produce pollution when they are driven
- they run quietly and smoothly
- electric motors are more efficient than diesel or petrol motors, so less energy is wasted running the engine
- renewable sources of energy can be used to generate the electricity, further reducing pollution.

Disadvantages of electric cars are:

- the battery has a limited range of about 100 miles
- recharging the battery takes up to an hour and it can be hard to find a suitable recharging point
- pollution is caused in generating the electricity needed to recharge the battery
- they tend to be small and do not have much power
- pedestrians do not always hear them coming.

Hybrid cars produce no pollution in heavy traffic

Improving electric vehicles

D4

Hybrid cars use more than one energy source. They are powered by electric batteries and also have fuel engines. They automatically use 'electric mode' for city driving, and change to 'fuel mode' for faster roads.

Advantages of hybrid cars compared with fully electric cars are:

- they have a greater range
- they are very efficient because the engine switches mode to suit the type of driving
- the battery recharges when the car is being driven in fuel mode
- there is less risk of being stranded by the battery running out of charge.

Disadvantages are:

- they are expensive because they are complicated to build
- the fuel engine produces pollution.

Think about

How would you decide if the extra cost of a car is worth the savings on fuel bills when it is used?

...hybrid car

Assessment and Grading criteria

To achieve a pass grade, my portfolio of evidence must show that I can:

Assessment Criteria	Description	✓
P1	Carry out an investigation into an application of the uses of motion	
P2	Carry out an investigation into an application of the uses of force	
P3	Carry out an investigation into an application of the uses of waves	
P4	Carry out an investigation into an application of the uses of electricity.	

Nuclear power station with two atomic reactors

To achieve a merit grade, my portfolio of evidence must show that I can:

Assessment Criteria	Description	✓
M1	Analyse the results of the investigation into the uses of motion	
M2	Analyse the results of the investigation into the uses of force	
M3	Analyse the results of the investigation into the uses of waves	
M4	Analyse the results of the investigation into the uses of electricity.	

To achieve a distinction grade, my portfolio of evidence must show that I can:

Assessment Criteria	Description	✓
D1	Evaluate the investigation into the uses of motion in our world, suggesting improvements to the real-life application	
D2	Evaluate the investigation into the uses of force in our world, suggesting improvements to the real-life application	
D3	Evaluate the investigation into the uses of waves in our world, suggesting improvements to the real-life application	
D4	Evaluate the investigation into the uses of electricity in our world, suggesting improvements to the real-life application.	

LO

Be able to investigate factors which contribute to healthy living

- Scientists have found a clear link between lifestyle, health and disease; people who lead unhealthy lifestyles are more likely to suffer from certain diseases

- The cost of work related mental illness made up a quarter of the UK's total sick bill in 2009

- Some peoples' diets can make them sick; people with Coeliac disease have intolerance to wheat and it can cause diarrhoea, stomach cramps and mouth ulcers

LO

Be able to investigate how some treatments are used when illness occurs

- The overuse of antibiotics for infections may be responsible for new bacteria that are resistant to antibiotics such as MRSA

- You cannot receive any type of blood if you need a blood transplant; you need blood that is complementary or a match to your own type

- Currently gene therapy is being developed for diseases like HIV and cancer and could provide an effective treatment in the future

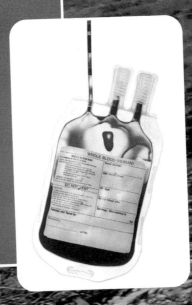

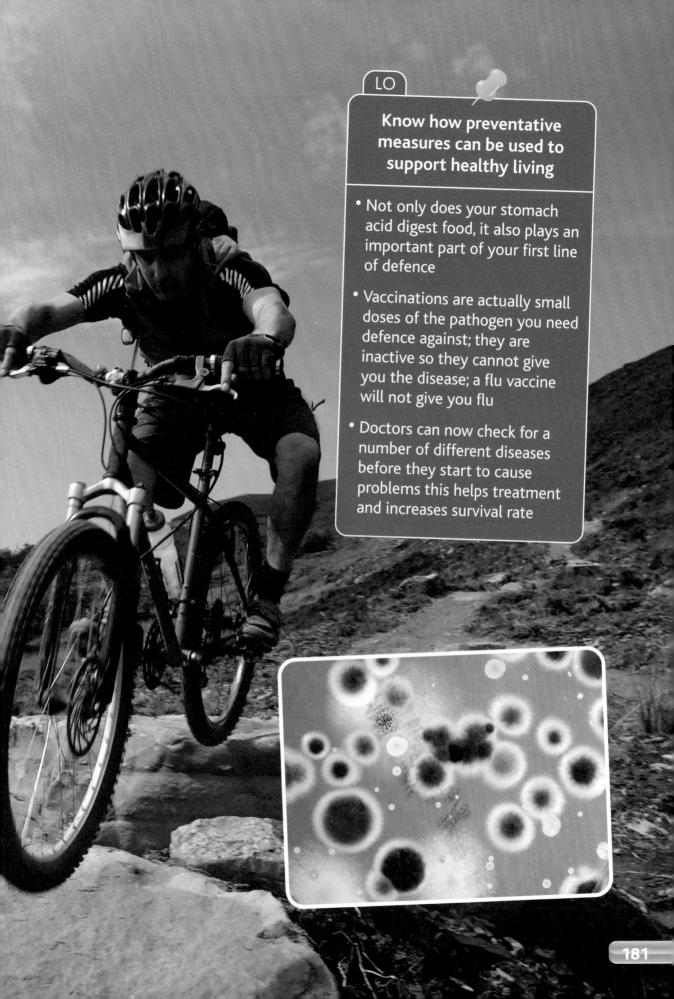

Know how preventative measures can be used to support healthy living

- Not only does your stomach acid digest food, it also plays an important part of your first line of defence

- Vaccinations are actually small doses of the pathogen you need defence against; they are inactive so they cannot give you the disease; a flu vaccine will not give you flu

- Doctors can now check for a number of different diseases before they start to cause problems this helps treatment and increases survival rate

Healthy living

Your assessment criteria:

P1 Assess the possible effects of diet on the functioning of the human body

P2 Design diet and exercise plans to promote healthy living

M1 Explain how the diet and exercise plan will affect the functioning of the human body

D1 Evaluate your exercise plans and justify the menus and activities chosen

Are you healthy?

Do you enjoy an evening on the sofa, watching TV, eating junk food and doing very little exercise? This may sound tempting and is common for some people. This type of lifestyle is proving to be a real concern. There are fears we are heading for an 'obesity epidemic'. This will result in more people becoming ill.

P1 P2 Factors that affect health

'Health' can be thought of as the general state of your body and mind. If someone is healthy then their mind and body are functioning well and they can live a full and active life. If someone is in a poor state of health, their body or mind or both are not working well.

Health is affected by a number of factors:

- **lifestyle** – the way we choose to live our lives
- **diet** – what we eat
- **exercise** – how active we are
- **physical illness** – things that can go wrong with our bodies, either caught (infectious) or developed (non-infectious)
- **mental illness** – things that can go wrong with our minds.

These factors are interrelated. For example, a lifestyle can be one that involves little exercise and therefore this can affect health. A poor diet can lead to a non-infectious physical illness such as heart disease. A lifestyle choice such as smoking can lead to a non-infectious disease such as lung cancer. Our health is in a constant state of change and the choices we make can directly affect it.

Know more

You learnt about infectious diseases in Unit 3. You can find out more in Unit 14.

Know more

Half of the world's population live on a diet of mainly rice.

Rice is a healthy food but rice alone cannot supply all the nutrients we need

M1

D1

Discuss

Are healthier people happier? What do you think?

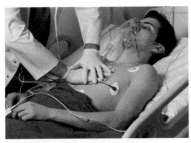

Even if you're not overweight, an inappropriate diet and lack of exercise can put pressure on your heart

Think about

Exercise doesn't have to mean going to the gym. Suggest how someone who is fairly inactive might increase the amount of exercise they take. How might a small increase in exercise affect the functioning of their body?

Think about

What would you do to assess someone's general state of health?

The benefits of a healthy lifestyle

A healthy lifestyle can have both **physiological effects** and **psychological effects**.

- If we control our diet and take exercise, our bodies become fitter. This can reduce the risk of some physiological conditions (physical illnesses) such as diabetes and heart disease.

- A healthy lifestyle in turn can affect our minds. We look better, we start to feel better about ourselves and this improves our moods. There is some evidence to suggest that this can help control psychological problems (mental illnesses) like depression.

Evaluating health

We may feel healthy, but can only decide how fit we really are by looking at our eating and exercise habits. It is easy not to notice how much we eat and drink, or what we are eating and drinking, and what the effects might be.

Some people need to structure their eating and exercise habits. For some this requires advice from dieticians, medical practitioners or fitness coaches. These experts help evaluate current lifestyles, design appropriate plans and review a person's progress towards their goals.

A fitness programme can help a person reach their health goals

...healthy lifestyle

Food and diet

Your assessment criteria:

P1 Assess the possible effects of diet on the functioning of the human body

P2 Design diet and exercise plans to promote healthy living

M1 Explain how the diet and exercise plan will affect the functioning of the human body

D1 Evaluate your exercise plans and justify the menus and activities chosen

Eating too much

Tom has over-indulged over the holiday period. He notices his trousers are tighter around the waist than they would normally be. He knows the reason – too much rich food, too little exercise. But he really didn't expect to put on so much weight. He wishes he could eat what he wants without any consequences.

P1
P2

We all need food

We all need food as our source of **nutrition**. It provides us with energy to carry out everyday activities, and essential **nutrients** for maintaining and repairing our bodies. If our meals over a period of time do not provide the right mix of nutrients in the right quantities, we can gain too much weight, lose too much weight, or damage our health in other ways.

The nutrients we need are shown in the table.

Nutrient	Why we need it
carbohydrate (sugar and starch)	This is our major energy source. The amount you need depends on the type of carbohydrate, who you are and how you live your life.
lipids (fats and oils)	These are energy sources and also have other functions in the body, such as making cell membranes.
protein	This is needed for growth and repair of the body. The amount you need depends on the type of protein, who you are and how you live your life.
vitamins and minerals	These are essential substances needed in very small amounts for the maintenance and effective working of the body.

Carbohydrates and fats are energy providers. Food with a lot of sugar or a lot of fat has a high calorific value. If we eat more than we need, the body stores the excess in the form of body fat and we put on weight. If we put on too much weight, we might find it difficult to move our bodies as easily, and extra pressure is put on the heart and lungs. Health problems can result.

Know more

*The amount of energy a food provides is called its **calorific value**. This is measured in kilocalories, kcal (often just called 'calories'), or kilojoules, kJ. We all need different amounts of energy from our food – see Unit 3.*

Know more

We need many different vitamins, for example vitamin C and vitamin B12, and several minerals such as magnesium and iron. Each one has a 'recommended daily amount' which you can find on food labels.

...healthy diet

M1

Think about

How might the dietary needs for a 60-year-old man compare to those of a 14-year-old boy? How might lifestyle and personal circumstances affect these dietary needs?

Think about

How would you explain the need for a low-calorie diet to an overweight person?

Are you on a diet?

There is often misunderstanding about the word 'diet'. Some people understand it to mean reducing what you eat in order to lose weight. Although the word can mean this, it actually means the food and drink that an individual regularly consumes. A diet can be made up of any food, and it can therefore be seen as 'good', 'poor', 'varied' or 'balanced'.

To maintain good health, a person's diet should be varied and balanced. A **balanced diet** provides all the nutrients needed, in the right proportions. Without a balanced diet there can be long-lasting effects on health. A person's nutritional and energy requirements vary according to their activity level, age, sex, muscle mass, health and environment.

If a person is regularly consuming too many calories per day, then a diet plan would aim to reduce the number of those calories. If followed, the body would respond by using some of its 'energy stores' (usually fat) and the person would lose some weight.

bread or cereal   fresh fruit and vegetables   water   oil

protein such as fish

A balanced diet is essential for good health

D1

Evaluating diet

When we evaluate what we eat, we need to look at the type of food we are eating.

Packaged food in the UK contains nutritional information on the pack. This tells us the calorific value and the quantities of each of the major nutrients in 100 g of the food. These data help us monitor our diets.

We can calculate the calories we take in by:

$$\text{number of calories} = \frac{1}{100} \times \begin{array}{l}\textbf{calorific value per 100 g} \times \\ \textbf{amount of food in grams}\end{array}$$

For example, if we eat 100 g of cake which has a calorific value of 379 kcal per 100 g, we would be consuming 379 kcals.

🔍 ...balanced diet

Diet and health

Super-skinny

Models are often employed because they are slim and look elegant in clothes. We assume they must be healthy because they are so slim, but model Luisel Ramos proved this assumption wrong. Luisel died at the age of 21 from a heart attack. She weighed only 44 kg. It is alleged her heart failure was a direct result of an eating disorder. Luisel was rumoured to have eaten only lettuce leaves and Diet Coke for a week before she died.

P1 P2

Getting the balance right

When a person eats too much it can be called **overnutrition**. This can lead to weight gain and in extreme cases can lead to **obesity**. There are a number of health conditions, such as heart disease, that are associated with being overweight or obese.

If a person eats too little they will suffer from **undernutrition**. They are not consuming enough calories and nutrients. The body responds by first breaking down its own fat and using it for energy. Weight loss occurs. After fat stores have been used up, the body may break down its other tissues, such as muscle and tissues in internal organs. This can lead to a number of different nutrient deficiencies and serious health problems, possibly death.

Eating disorders

In the UK and other wealthy nations the choice of what we eat is under our own control. For some people this control goes wrong and they can develop **eating disorders**.

Know more

Eating disorders are around ten times more common in girls and women than in boys and men. They affect seven teenage girls in every 1000, but only one teenage boy in every 1000.

🔍 ...undernutrition

P1
P2

The eating disorders **anorexia nervosa** and **bulimia** are both serious psychological conditions, leading to severe physical problems, and in extreme cases death.

The anorexia nervosa sufferer has a greatly reduced appetite and obsessive control of what they eat, or even a complete avoidance of food.

The bulimia sufferer alternates between eating excessive amounts of food (bingeing) and making themselves sick, or using laxatives (purging), in order to maintain a chosen weight.

Food allergies

In some people certain types of food cause an **allergic reaction**. This may be mild or extreme. Sometimes, if appropriate treatment is not obtained, it can be fatal. The most well known food allergies are to peanut and egg.

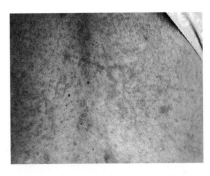

This rash has been caused by an allergic reaction to seafood

M1

Discuss

What power does the mind have over a successful diet plan? What could be the potential consequences of placing an overweight person on a calorie-restricted diet?

Dealing with undernutrition

Undernutrition can occur for a number of reasons. In today's society it can still occur because of poverty, or when natural disasters such as flood or drought ruin food supplies. In parts of society with plenty of food available, undernutrition can occur because of psychological eating disorders.

The treatment depends on the circumstances. In disaster-stricken areas, short-term measures may include distributing high-protein biscuits. Long-term solutions always mean looking at the underlying cause. People with a psychological disorder need long-term treatment to change the way they view themselves and their attitude to food.

A high-protein shake is sometimes recommended for people who need to put on body mass

D1

Crash diets

It is tempting to go to extremes when dealing with a diet. Some people believe that a diet to lose weight requires starvation – a crash diet. People may think that to put on weight they need to eat more high-fat and sugary foods. These methods, although they might lead to weight loss or gain in the short term, can cause problems in the long term. The key to any successful diet plan is *balance*.

A crash diet is unlikely to be the right kind of diet plan for an obese person

...eating disorders

Fit and healthy

Your assessment criteria:

P1 Assess the possible effects of diet on the functioning of the human body

P2 Design diet and exercise plans to promote healthy living

M1 Explain how the diet and exercise plan will affect the functioning of the human body

D1 Evaluate your exercise plans and justify the menus and activities chosen

Looking good, feeling good?

A very muscular body is something that some men aspire to. More visible muscles are often seen as attractive and 'fit'. But this is not always the case. Appearances can be deceptive. A person who looks good might not be fully fit and healthy.

P1 P2

Exercise for fitness

Exercise is an important part of a healthy lifestyle. Someone can look good but may not include regular exercise in their lifestyle. If this is the case then they may be unfit even though they look quite good.

Fitness is the ability to do what your body is required to do. It can include agility, stamina, balance, speed and flexibility. Fitness is different for different people.

The amount of exercise needed to maintain fitness depends on the fitness level of the person and the goals set. For example, if you were an Olympic marathon runner, jogging one mile in nine minutes would count as only a **mild intensity** activity. For many people, though, it would be a **high intensity** activity. Experts recommend that, for most people, about 30 minutes of **moderate intensity** exercise five times a week is needed to be fit.

The fitness level you can achieve depends on your current physical state. Some people have restricted **mobility** (ability to move). These may be very obese people, or those recovering from surgery, or those with chronic conditions that affect mobility, such as degenerative joint and muscle diseases (like arthritis). These people would need to work on exercises that gradually increase their range of mobility and flexibility. Small gains in fitness could have significant impact on their lifestyles.

For others, the goal might be regaining a previous fitness level (regaining their form). Competitive sportspeople often suffer from injury, as any strenuous exercise has the potential to damage joints, break or dislocate bones, and tear muscles and ligaments. After recovering from the injury, they would need a training programme designed to achieve their former fitness level.

Know more

*Being **agile** means you can move in sudden bursts in different directions, like running in zig-zags down a football pitch. Being **flexible** means you have a wide range of movement around your joints and may be able to 'do the splits'.*

If you can't do this, does it mean you're not fit?

Know more

It is estimated that only 37% of men and 24% of women take enough exercise.

🔍 ...exercise and health

Personal computers can take people through fitness programmes

Think about

Are you fit? List the activities that you do which contribute to your fitness.

Discuss

What might be a realistic goal for someone who is extremely unfit?

The right type of exercise

Different activities in a health and fitness plan have different effects on the functioning of the body. Some examples are shown in the table.

Activity	Effect on the body
aerobic exercise (such as running, cycling)	Improves speed and agility, increases lung capacity and **cardiovascular fitness** (how fit your heart and circulatory system is), reduces stress and can lead to weight loss.
yoga	Improves balance and flexibility and can tone up parts of the body.
weights	Builds up and tones muscles, can change body shape and appearance.
jogging	Increases lung capacity and cardiovascular fitness, reduces stress and can lead to weight loss.
swimming	Improves speed, stamina, cardiovascular fitness and tones the body. Can lead to weight loss and reduces stress.

M1

Evaluating fitness

Different **physiological measurements** can be made to establish how fit someone is:

- **resting pulse rate** – the normal pulse rate when not exercising
- **recovery time** – how long it takes for the pulse and breathing rate to return to normal after exercise
- **body mass index (BMI)** – shown by this ratio:

$$\frac{\text{weight}}{(\text{height})^2}$$

- percentage of body fat
- waist circumference
- waist/hip ratio.

There is a range of devices on the market for making these measurements.

An evaluation of fitness should not just rely on one method – it is better to use a combination of methods.

D1

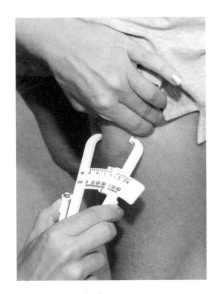

Measuring body fat

🔍 ...exercise and pulse rate

Healthy heart

'I had a heart attack'

Lee wasn't overweight, but he had high blood pressure and high levels of cholesterol in his blood. This combination and his lack of exercise led to the day where his struggling heart couldn't take it any more. He had a heart attack. He survived and will be looking after his health a lot more in the future.

Your assessment criteria:

P1 Assess the possible effects of diet on the functioning of the human body

P2 Design diet and exercise plans to promote healthy living

M1 Explain how the diet and exercise plan will affect the functioning of the human body

D1 Evaluate your exercise plans and justify the menus and activities chosen

P1
P2

Problems with the circulatory system

The circulatory system (or cardiovascular system) consists of the heart, blood vessels and blood. One of its jobs is to transport essential substances around the body. It transports nutrients from food to cells, and waste products away from cells.

In a healthy circulatory system the heart and vessels work efficiently. But things can go wrong. A number of conditions can develop that are directly related to diet and lifestyle:

- **heart disease**

- **arteriosclerosis** (blocked artery)

- **high blood pressure**

- **stroke** (sudden change in blood flow to the brain).

Factors such as high blood pressure, an unbalanced diet, smoking, being overweight, having an inactive lifestyle, family history and stress can all contribute to an unhealthy circulatory system. The heart will have to work harder, which can mean that the muscle becomes overworked and tired, and it can stop working as well.

What is cholesterol?

Cholesterol is a fatty substance that is made in the body by the liver and found naturally in the blood. Some cholesterol is needed to help cells work well. It also produces vital chemicals such as vitamin D and some hormones.

Regularly eating meals like these could increase blood cholesterol level

...cholesterol valves

Know more

There are two types of cholesterol:

- HDL (high density lipoprotein), sometimes known as 'good' cholesterol
- LDL (low density lipoprotein), sometimes known as 'bad' cholesterol.

It is a high level of LDL in the blood that is a problem.

Our diet can affect the amount of cholesterol in our blood. Foods that are high in **saturated fat** – including red meat, sausages, cheese, butter, cream, pastry, cakes and biscuits can increase blood cholesterol level.

If there is too much cholesterol in the blood it gets deposited on the walls of the blood vessels, builds up as plaque and can restrict the blood flow.

Plaque build-up can block blood vessels

Think about

Plan a low cholesterol diet for a week.

Keeping the circulatory system working well

A balanced diet that is low in saturated fat will mean there is less chance of plaque building up in the blood vessels. There is therefore less chance of stroke or heart disease.

Regular exercise, especially aerobic exercise, can:

- strengthen the cardiovascular system
- improve the circulation
- lower the blood pressure
- improve muscle tone and strength, bone strength and joint flexibility
- help reduce body fat and reach or maintain a healthy weight.

Research

What evidence is there that cholesterol-reducing spreads and drinks actually work?

Evaluating the health of the circulatory system

- The level of blood cholesterol can be monitored by a simple blood test.
- Blood pressure can be easily checked by a nurse or doctor.
- Resting pulse rate (usually lower in someone with a healthy circulatory system) can be easily measured by counting.
- Recovery time after exercise can be easily timed (people with a healthy circulatory system tend to return more quickly to a normal pulse rate and breathing rate).

Taking a pulse rate is an important way of monitoring the health of the circulatory system

Planning for health and fitness

Getting excited about health

Too much rich food, too little exercise, and all of a sudden we feel unhealthy. The good news is that it is never too late to do something about it. By following a healthy living plan and establishing good habits you can return to health. Healthy habits and exercise can be fun – even cycling with your family once a week can have a positive effect on your health.

P1 P2 The components of a health plan

To design a health and fitness plan we need to first evaluate how healthy a person is. We can do this in a number of ways. One type of evaluation is a **physiological investigation**. This consists of taking a number of measurements for the person, as you learnt earlier in this unit.

We also need to make a **lifestyle evaluation**. This is a series of questions relating to the way the person lives their life. It could include questions on:

- diet and eating habits
- sleeping habits
- exercise habits
- job and other activities
- smoking and alcohol intake
- attitude towards food and exercise.

Once the data has been collected it is analysed, and the areas that need to be improved are decided on.

...health and fitness plans

P1
P2

Short-term goals may be set, with a suitable time frame.

'I would like to lose 3 kg in 2 months.'

For some people it is appropriate to set **long-term goals**.

'I need to improve my fitness to run a half-marathon within a year.'

A specific health and fitness programme can be developed that is tailor-made for the individual. It should address all the issues raised in the evaluation.

FITNESS PLAN

Name _____

Age _____

Height _____

Injuries/health conditions _____

Aims _____

A fitness plan needs to take into account key measurements

The benefits of a health and fitness plan

M1

A diet and exercise plan can have beneficial effects on the health and fitness of a person. However, what is appropriate for one person may not be for another. It depends on the initial level of fitness and on the goals.

Keeping to a healthy diet can help maintain a suitable weight, with body fat in the appropriate range, and reduce cholesterol levels.

Keeping to a fitness plan can benefit the heart and circulatory system, can help maintain a suitable weight and percentage of body fat, and for some people it can increase movement and flexibility and improve mental health.

Good diet and exercise regimes can help reduce the risk of heart disease.

Fish oils are very beneficial to our health

Research

What alternatives to fish oil could you find for vegetarians or people who do not like fish, which would provide the same nutritional benefits?

Think about

How would you assess the impact of a diet and exercise plan on the habits and health of a person? How could a plan be amended if a particular exercise was not working for the person?

Making the plan fit for purpose

D1

A diet and fitness plan is only as good as the results it delivers. To evaluate a plan you need to look at the goals of the plan and establish whether they have been reached. For example, the person's goal might be to lose weight in order to reduce BMI. The person's BMI – weight divided by (height)^2 – would be calculated before, during and after the programme to see if the goal has been met.

Preventing illness and disease

Ill all the time?

Imagine if we caught every virus around. We'd be ill all the time. We couldn't go to school, go to work, go out to socialise, we may not even be able to go out at all. Our lives depend on us being healthy and so we have a well-tuned system for preventing illness.

P3

The first line of defence

Illnesses can have either a **physical cause**, such as injury, or a **biological cause**, such as a **pathogen** infecting the body.

A broken arm is a physical cause

The human body is designed to work efficiently. It has sophisticated systems to help prevent illness. These involve different 'lines of defence':

- **physical barriers** help to stop pathogens entering the body
- **chemical barriers** deal with invaders if they do get in.

Pathogens can get into the body through the skin, the eyes, the ears, and through external respiratory organs (mouth and nose), digestive organs (mouth and anus), urinary organs (urethra) and reproductive organs (vagina and penis). For example, pathogens can invade through damaged skin, we can **ingest** them with our food, or we can **inhale** them through the nose.

Know more

A pathogen is anything that can cause disease. It is usually a bacteria or a virus. These are types of micro-organism, or microbe. See Unit 3.

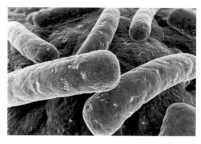

A sore throat is caused by bacteria – a biological cause

Know more

By 'chemical barrier' we mean a substance in the body that can kill, or limit the effect of a pathogen.

...first line of defence against disease

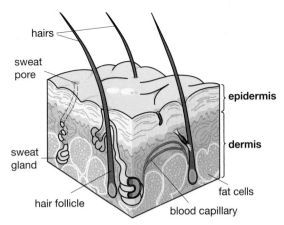

The skin is our most important barrier against pathogens. It is waterproof and slightly acidic

M2

Discuss

If we have all these barriers, why do we still get ill?

The barriers at work

- When skin is broken by a cut, it rapidly blocks the break and any cut blood vessels, forming a scab. This prevents the entry of microbes and allows the skin time to heal.

- In the respiratory system we have hairs in our nose that capture large particles, and **mucus**, a sticky substance that traps dirt and large pathogens. If we get infected we often make more mucus. This is a symptom of a cold. In our lungs we have cilia (hair-like projections that move) which help to remove mucus and pathogens from the lungs.

- In the digestive system, pathogens that have been swallowed may be killed by **stomach acid**. Special chemicals called enzymes also help to destroy pathogens. The inner lining of the gut produces mucus that can trap invading pathogens. The intestines have so-called 'good bacteria' that help prevent 'bad bacteria' taking over.

- In the **urinary system**, the flow of urine flushes bacteria out of the bladder and urethra.

When these barriers fail, we have a second line of defence known as the immune system. This will be the subject of the next section.

Research

What would happen if we didn't have these physical and chemical barriers as a first line of defence? Find out about burn victims whose skin has been severely damaged.

Think about

How can medicine help if there is too much damage to the skin (too large a cut, graze or burn)?

🔍 ...protective role of skin

The immune system

Chickenpox

Laura has chickenpox. Many children get this disease. It is caused by a virus. It is not a pleasant infection because it causes a very itchy rash that turns to blisters, as well as flu-like symptoms.

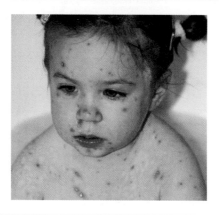

Your assessment criteria:

P3 Outline how the immune system defends the human body

M2 Describe the action of each component of the immune system

M3 Describe the changes in the human body following a vaccination

D2 Evaluate the effectiveness of vaccination and screening programmes

P3 The second line of defence

If your body has been infected by a foreign substance or cell (pathogen), it responds to it and defends itself using the immune system. This fights the infection and can sometimes prevent it from re-occurring.

The immune system is made up of white blood cells which search out the infection, identify it and fight it. White blood cells are made in the bone marrow. They move around the body in the blood to where they are needed.

There are different types of white blood cell. Each has a specific way of working, depending on the threat.

- **Phagocytes** seek out and destroy pathogens or other foreign substances by engulfing (eating) them.

- **Lymphocytes** work in a different way. They identify particular pathogens and produce substances called antibodies. This enables them to destroy or de-activate the pathogens.

Know more

A drop of blood can contain anywhere from 7000 to 25 000 white blood cells.

Know more

If your immune system is able to recognise a particular pathogen and fight it so that you do not get the disease, you have **immunity** *against that disease.*

M2 Phagocytes and lymphocytes

Phagocytes engulf any types of pathogen or foreign substance and then digest them using enzymes.

Lymphocytes identify and fight against pathogens or abnormal body cells. They can stick to a pathogen or abnormal cell, making it more likely to be destroyed. They produce antibodies to help de-activate specific pathogens, and can also produce substances that counteract poisons produced by some pathogens.

Research

Which conditions directly affect a person's immune system?

 ...immune system

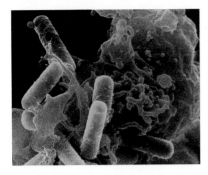

A phagocyte (orange) engulfing bacteria (blue)

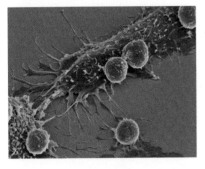

Lymphocytes (pink) attacking a cancer cell

Think about

What is the difference between active and passive immunity?

Research

For what diseases are there vaccination programmes in the UK? Try to find some data about how effective they are.

Active and passive immunity

M3

After an immune response to a pathogen, the body 'remembers' the pathogen and has a defence ready for future attacks. This is known as **natural active immunity**.

When you are vaccinated (see Unit 3), you are given an inactive form of the pathogen that stimulates an immune response. Antibodies that work against that pathogen are produced. This leads to immunity known as **artificial active immunity**.

In both of these cases, the immune response can take a while to develop. If a person is infected with tetanus, this can be fatal because the toxins produced can quickly kill. The person needs to be given a readymade dose of antibodies against the infection. This gives immediate immunity and is called **artificial passive immunity**. A mother can give passive immunity to her baby naturally, via the placenta or her milk.

Effectiveness of vaccination

D2

The effectiveness of a vaccination programme is measured by how well it prevents a population from getting the disease. It depends on what proportion of the population is vaccinated. Everyone in the UK is offered the polio vaccine. Babies are routinely vaccinated. Polio has disappeared here. But in some countries in Africa it is still a major disease because the vaccine cannot be offered to all. Even where it is available, not everyone takes it up because of health fears associated with vaccination.

Vaccination has one big success story. In 1979 the World Health Organisation officially recognised that smallpox, a killer disease, had been wiped off the planet.

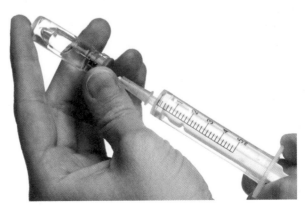

Preparing to give a vaccination

...vaccinations

Screening for disease

Jade's fight for life

Jade Goody died from cervical cancer in 2009. If her cancer had been detected earlier she might have survived. Her well publicised fight against cervical cancer helped to raise awareness of cervical cancer screening.

Your assessment criteria:

P4 Identify the role of specific health screening programmes

M4 Using secondary data, carry out an investigation into the effectiveness of different kinds of medical treatment in the control of health

D2 Evaluate the effectiveness of vaccination and screening programmes

P4

Preventative measures

Scientists have developed **screening programmes** which help detect any abnormalities which might indicate the early stage of a disease or other condition. Screening is usually targeted at specific 'at-risk' groups, such as women of a particular age range.

Current screening programmes include the following.

- **Breast screening** is a method of detecting breast cancer at a very early stage. The first step involves an x-ray of each breast, called a **mammogram**. This can show small changes in breast tissue which are too small to be felt either by the woman herself or by a doctor.

- **Cervical screening** is a method of detecting early abnormalities in a woman's cervix (the neck of the womb) which, if left untreated, could lead to cancer.

- **Antenatal screening** is a way of assessing whether an unborn baby (foetus) could develop or has developed an abnormality during pregnancy, for example spina bifida or Down's syndrome.

- **Newborn** or **neonatal screening** helps to identify babies who may have rare but serious conditions. Babies in the UK are usually screened for specified conditions including sickle cell disorders and cystic fibrosis.

- **Vascular screening** usually involves a series of health checks such as height, weight, cholesterol level, blood pressure, and family history, to establish the risk of cardiovascular (heart or circulatory) disease.

Know more

Around 1.5 million women are screened in the UK each year for breast cancer.

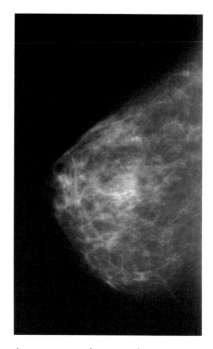

A mammogram is a screening test to detect breast cancer in its early stages

⌕ ...health screening, NHS cancer screening

P4

Discuss

When funding is limited, is it better to spend money on treating conditions, or on screening for conditions?

Screening for disease is not compulsory but is recommended by health professionals.

Some types of screening, such as ante- and neonatal checks, require appointments in hospitals for specialist tests. Other screening, for example for cardiovascular disease, can be achieved by regular health checks with the GP. Some screening can be done by the individual at home, for example self-examination to check for breast or testicular lumps.

Neonatal screening can help detect some rare and serious conditions

M4

How effective is screening?

Screening can be very effective in the reduction of the number of cases of certain diseases. About 1 in 20 people in the UK will develop bowel cancer during their lifetime. It is the third most common cancer in the UK, and the second leading cause of cancer deaths. Over 16 000 people die from it each year. The NHS states that regular bowel cancer screening has been shown to reduce the risk of dying from bowel cancer by 16%. Men and women aged 60 to 69 are now routinely screened.

Think about

What factors might be considered by an individual when deciding whether to be screened, or whether they want their child to be screened, for a particular disease?

Research

Find other diseases that are now screened for in the UK, and their 'target population'.

D2

Evaluating effectiveness

An evaluation of a screening programme relies on collecting data as evidence of its effectiveness. Effectiveness can be measured in different ways:

- the number of patients being identified and treated for the disease
- the **coverage** of the screening programme (the percentage of the target population screened)
- the reduction in serious cases of (or deaths from) the disease
- the cost-effectiveness.

Treatment options

Your assessment criteria:

P5 Carry out an investigation into the effects of antibiotics

Practical

P6 Describe what gene therapy is, giving examples of diseases and conditions associated with it

M4 Using secondary data, carry out an investigation into the effectiveness of different kinds of medical treatment in the control of health

D3 Evaluate the use of different kinds of medical treatments, justifying your opinions

Treatment in the Middle Ages

For thousands of years, doctors relied heavily on one kind of treatment for many conditions. It didn't matter whether someone had a persistent headache or heart disease: the 'remedy' was bloodletting. Doctors would routinely cut open a blood vessel and let out blood. Fortunately our knowledge of how the human body works has increased, and medicine today has a vast range of treatments available.

P5 Physical and chemical therapies

Treatment options depend on whether the condition can be cured, or if the aim is to control it. Some treatments control the symptoms, such as pain or discomfort, but do not cure the condition. Treatments can be thought of as either physical therapies or chemical therapies.

- A **physical therapy** is a type of treatment that does something physically to the body. An example of a physical therapy is surgery. A physical therapy is particularly useful when specific parts of the body are not functioning well.

- A **chemical therapy** is a type of treatment that uses chemicals (drugs) to treat the cause and/or the effects of an illness. For example, antibiotics are drugs that can kill bacteria that have invaded the body.

A **replacement therapy** is a physical therapy that involves replacing parts of the body that are not working well. Examples of this are **transplants** and blood transfusion.

Know more

There are different blood groups in humans, and blood from one person cannot be transfused into a person who has a non-compatible blood group. This could be fatal.

...blood transfusion ...organ transplants

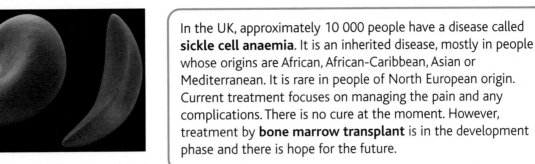

In the UK, approximately 10 000 people have a disease called **sickle cell anaemia**. It is an inherited disease, mostly in people whose origins are African, African-Caribbean, Asian or Mediterranean. It is rare in people of North European origin. Current treatment focuses on managing the pain and any complications. There is no cure at the moment. However, treatment by **bone marrow transplant** is in the development phase and there is hope for the future.

P5

Sickle cell anaemia affects the red blood cells; this disease makes them sickle-shaped as on the right, and alters their function so that they cannot carry much oxygen

Know more

Gene therapy will be looked at in detail at the end of this unit.

Research

Find out where stem cells are obtained from and explain why their use is controversial. What other conditions can they potentially treat?

Discuss

Should the government fund stem cell research?

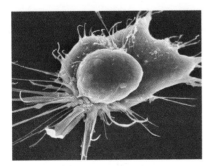

The use of this simple stem cell has medical potential but is hotly debated

Gene therapy

P6

Gene therapy is a sophisticated treatment that is currently being researched. It can potentially be used to treat serious **genetic disorders** – inherited diseases passed from parents to children. An example is cystic fibrosis (see Unit 3).

Stem cell transplants

M4

Stem cells are body cells that can develop into other, specialised cells that the body needs. The stem cells in bone marrow develop into our immune system. The disease leukaemia destroys the immune system. If stem cells are transplanted into someone with leukaemia it can treat the disease. Stem cell treatment has proven effective in some cases but is complex and depends on multiple factors.

Evaluating treatment options

D3

Treatments can be evaluated from a number of different perspectives.

Medical and scientific: Is it working? What are the benefits? What is the evidence? What are the risks?

Financial: How much does it cost? Are the benefits of the treatment worth this expenditure?

Ethical: Are there any reasons why the treatment could be considered wrong? Is the treatment essential, whatever the cost?

Social: Does the treatment benefit society as a whole? Will the treatment take too much money away from other important areas?

Religious: Do religious beliefs reject the treatment?

...stem cell research

Antibiotics

Your assessment criteria:

P5 Carry out an investigation into the effects of antibiotics

Practical

M4 Using secondary data, carry out an investigation into the effectiveness of different kinds of medical treatment in the control of health

D3 Evaluate the use of different kinds of medical treatments, justifying your opinions

Arguing about antibiotics

On a visit to the doctor's surgery with a really bad cold, Julia had a disagreement with her doctor. Julia wanted an antibiotic that would cure her of her cold, but the doctor said he couldn't give her anything. He told her that antibiotics only work on bacterial infections, not on viruses.

P5

The history of antibiotics

An antibiotic is a substance that kills or slows the growth of microbes, especially bacteria. It is made naturally by other microbes.

Antibiotic substances were discovered long ago, when the ancient Egyptians and Chinese used mould to treat infections. The breakthrough in medical treatment came when the effects of one particular organism were discovered in 1929. This was 'penicillium', discovered by Alexander Fleming. It was only in 1940 that antibiotic drugs were developed and used widely.

Types of antibiotic

There are many types of antibiotic. Each works slightly differently and acts on different types of bacteria. Some antibiotics are very good at fighting a wide range of bacteria. These are called **wide spectrum antibiotics**. Others are effective against only a few types of bacteria. These are called **narrow spectrum antibiotics**.

Some antibiotic names you may have heard of are:

Penicillin   Amoxicillin   Tetracycline   Erythromycin

Antibiotics are used to treat bacterial infections, including throat infections, chest infections, pneumonia, syphilis and meningitis. They work either by binding to the **cell membrane** of the bacteria and destroying it, or by stopping it **replicating**.

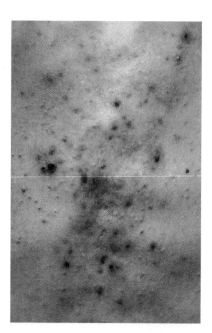

Acne is sometimes treatable with an antibiotic

Know more

Antibiotics cannot tell the difference between 'good' and 'bad' bacteria, so they can get rid of some of the bacteria you actually need.

...antibiotics

Think about

Why is it not a good idea to take only some of the antibiotic medicine?

Are antibiotics effective?

Antibiotics have been highly successful worldwide in treating a wide range of dangerous and potentially fatal diseases. Before Penicillin was developed, bacterial infections were a major cause of death. In 1900, the three leading causes of death were pneumonia, tuberculosis (TB), and diarrhoea and sickness. These caused one third of all deaths. All of these can now be effectively treated with antibiotics.

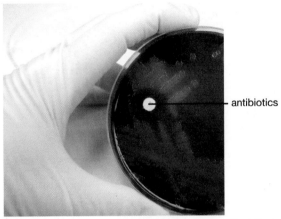

— antibiotics

The disc in this plate contains antibiotics, which have killed the bacteria around it

Antibiotics only work if the full amount of medicine prescribed is taken. Sometimes they can have side-effects or even cause allergic reactions.

A further major disadvantage is that inappropriate use and over-prescribing of antibiotics has helped to create strains of bacteria that are **resistant** to them. These so called 'super-bugs' are responsible for many infections that are caught in hospital. The most well known one is MRSA. Worryingly, a new antibiotic-resistant form of tuberculosis is now causing concern.

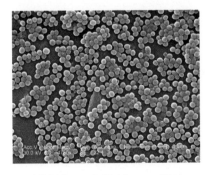

MRSA bacteria highly magnified

Discuss

Can we justify the continued use of antibiotics?

Evaluating antibiotic use

So how should we use antibiotics in the future? This will really depend on demand and if they remain effective. Antibiotic resistance is increasing. If doctors prescribe antibiotics less often, and if patients always complete the course of medicine prescribed, the development of resistance may slow down. It's not possible to stop resistance completely, but slowing down its spread buys us some time to develop new types of antibiotics.

🔍 ...antibiotic resistance

Gene therapy

Your assessment criteria:

P6 Describe what gene therapy is, giving examples of diseases and conditions associated with it

M4 Using secondary data, carry out an investigation into the effectiveness of different kinds of medical treatment in the control of health

D3 Evaluate the use of different kinds of medical treatments, justifying your opinions

A royal disease?

Queen Victoria passed on to her children the chance of getting the 'royal disease' – now known as haemophilia. This is a serious genetic disease that affects the clotting of the blood.

P6

Genetic diseases

When an egg is fertilised by a sperm, the new baby will have genes from the mother and father. These normally instruct the body how to grow and function correctly.

However, sometimes these genes are **mutated** (changed or not copied correctly) and this can have a devastating effect on the growing baby.

Huntington's disease is a condition that is caused by inheriting a mutated gene from parents. You cannot catch a genetic disease like this – you either have a mutated gene or you don't.

Other genetic diseases include haemophilia, sickle cell anaemia and **muscular dystrophy**.

At present there is little to be done to cure genetic diseases. Current medicine relies on controlling the symptoms and some of the associated conditions. However, there could be new hope on the horizon. Proposed new treatments for genetic conditions involve inserting new, correct genes into the ill person's cells. It is done using a **vector**, usually a virus, to carry the gene into a cell's nucleus. This is called gene therapy. The processes involved are shown in the diagram at the top of the opposite page.

Gene therapy is still in an experimental phase but has been successful in the treatment of some conditions.

Know more

If you have forgotten about genes, look back at Unit 3.

Know more

The first gene therapy trial was carried out in 1990. It is estimated that about 100 new trials are started every year, because there is such a large number of illnesses that it could potentially cure.

🔍 ...genetic diseases

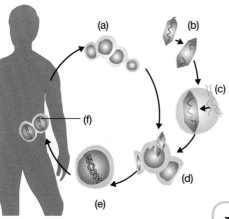

(a) Cells are removed from patient

(b) In the lab a virus is altered so it cannot reproduce

(c) A correct gene is inserted into the virus

(d) The altered virus is mixed with cells from the patient

(e) The cells become genetically altered

(f) The altered cells are injected into the patient

How gene therapy works

The future of gene therapy

M4

No-one yet knows if gene therapy is going to be generally effective. It has been effective in trials and in some research situations, and is currently being trialled in many areas of treatment, including:

- cystic fibrosis (see Unit 3)
- some types of cancer
- inherited blindness
- **thalassaemia** (an inherited disease where a damaged version of blood protein is made that leads to anaemia).

Before gene therapy can be used widely, a thorough investigation is needed of possible side-effects. These might include unwanted immune responses which could make the patient worse.

It is a controversial treatment because it is still in its developmental phase, its efficacy has been called into doubt, and it raises a number of ethical questions. Yet it remains a possible treatment with huge potential.

Think about

In a trial, gene therapy has proved effective in treating blindness in mice. Could it do the same in humans?

Evaluating gene therapy

D3

To evaluate this treatment we need to appreciate both the pros and cons.

For gene therapy ...	Against gene therapy ...
Gene therapy could change the lives of patients and families if it works.	Gene therapy is still in its experimental phase.
It gives some hope for people who have exhausted all other treatment options.	The trials are very expensive.
	It could have harmful side-effects.

Discuss

Gene therapy is not yet proven to work. Should it even be considered by people who have a genetic disorder?

...gene therapy

To achieve a pass grade, my portfolio of evidence must show that I can:

Assessment Criteria	Description	✓
P1	Assess the possible effects of diet on the functioning of the human body	
P2	Design diet and exercise plans to promote healthy living	
P3	Outline how the immune system defends the human body	
P4	Identify the role of specific health screening programmes	
P5	Carry out an investigation into the effects of antibiotics	
P6	Describe what gene therapy is, giving examples of diseases and conditions associated with it.	

Fruit and vegetables essential to our health

To achieve a merit grade, my portfolio of evidence must show that I can:

Assessment Criteria	Description	✓
M1	Explain how the diet and exercise plan will affect the functioning of the human body	
M2	Describe the action of each component of the immune system	
M3	Describe the changes in the human body following a vaccination	
M4	Using secondary data, carry out an investigation into the effectiveness of different kinds of medical treatment in the control of health.	

To achieve a distinction grade, my portfolio of evidence must show that I can:

Assessment Criteria	Description	✓
D1	Evaluate your exercise plans and justify the menus and activities chosen	
D2	Evaluate the effectiveness of vaccination and screening programmes	
D3	Evaluate the use of different kinds of medical treatments, justifying your opinions.	

Unit 10 **The Living Body**

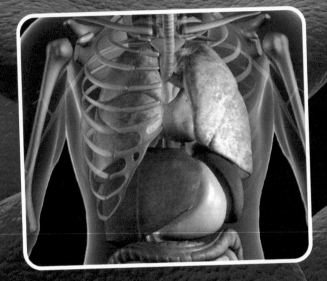

LO

Be able to investigate individual body systems, relating their structure and functions to their role in maintaining health

- In the 1950s it was common to see hospitals full of patients inside iron lungs because they were paralysed after catching polio

- A living person can donate part of their liver to someone with liver failure

- On a very hot day your urine becomes much darker in colour because you lose more water from sweating

- Your body contains 25 billion red blood cells; end-to-end, the blood cells would stretch four and a half times round the world

LO

Know the role of enzymes as catalysts

- 75% of humans can't digest the sugar in milk because they don't have the enzyme lactase in their digestive system

- Without enzymes we would not have cheese, yoghurt, and alcohol

LO

Know the structure and functions of the human reproductive system

- Women who live in the same house often find their monthly periods happen at the same time

- The contraceptive pill was first used in the 1960s, before that (and even now) the best way to avoid pregnancy was by not having sex

LO

Know how the nervous and endocrine systems work to coordinate the body systems

- To be good at sport you need excellent hand-eye co-ordination; your nervous system needs work efficiently

- Adrenaline targets more organs than most hormones, including your heart, stomach and eyes

- During adolescence your body releases more sex hormones which affect your body in all sorts of ways

Enzymes

Blood on your shirt again!

Paul has been playing football and was hit on the nose with the ball. There is now blood all over his shirt. His mum is not going to be pleased.

Luckily the blood stains can be broken down using a biological washing detergent.

P1 What are enzymes?

Enzymes are biological catalysts. They speed up reactions.

Biological washing detergents and stain removers contain enzymes that clean your clothes. The enzymes break down the stains so that they dissolve in the water.

These same enzymes are found in your digestive system. Your body needs them to speed up the digestion of food. If there were no enzymes, the nutrients would not be able to dissolve and pass into your blood.

Enzymes are also used to make some of the food we eat. Cheese can be made from sour milk. When milk turns sour it separates into a solid and a liquid. The solid part is called curds. It is the curds that form cheese. The chemical reaction that turns milk sour is slow. An enzyme called rennin speeds up the reaction so the cheese is made a lot faster.

Enzymes found in yeast are very important. The enzymes are used to turn sugar into alcohol and carbon dioxide. The alcohol can be turned into beer and wine. The carbon dioxide is used to make bread rise.

Know more

If you chew bread for long enough it starts to taste sweet. This is because enzymes in your mouth break down the starch in the bread. The starch becomes sugar.

Know more

Yeast is a living micro-organism. The enzymes speed up respiration in the yeast cells.

M1 Enzymes and rate of reaction

Enzymes are needed to catalyse many reactions in your body. The reactions start with molecules called **substrates**. These are changed in the reaction into **product** molecules.

 ...effect of temperature on enzymes

M1

Research

Try to find out which enzymes are in washing detergents. At what temperature do they work best? You could start your search by looking on a packet or bottle of washing detergent.

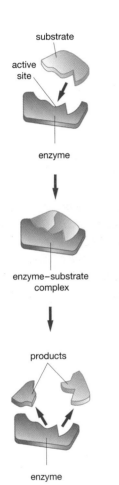

substrate

active site

enzyme

enzyme–substrate complex

products

enzyme

The 'lock and key' theory of enzymes

Think about

Use the science to explain the shape of the graph above. Can you draw a similar graph to show the effect of temperature on the rate of reaction?

Each enzyme is specific to one substrate. Fat is one example of a substrate. The enzyme lipase breaks fat down into two products, glycerol and fatty acid.

Enzymes can work faster if there is a higher concentration of substrate.

Enzyme-controlled reactions are also affected by the surrounding conditions. The **rate of reaction** depends on:

- pH, and
- temperature.

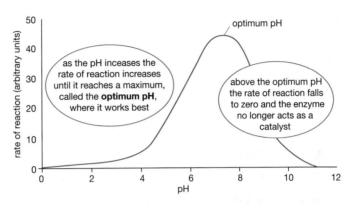

as the pH inceases the rate of reaction increases until it reaches a maximum, called the **optimum pH**, where it works best

above the optimum pH the rate of reaction falls to zero and the enzyme no longer acts as a catalyst

Graph to show how pH affects the rate of an enzyme-catalysed reaction

Each enzyme has a different shape. Within this shape is a structure called an **active site**. This is where the substrate joins to the enzyme. If the shape of an enzyme changes, the substrate will not fit into the active site. The enzyme can then no longer catalyse the reaction. The enzyme has become **denatured**.

D1

Analysing enzyme reactions

The effect of pH

Every enzyme has an optimum pH at which it works most efficiently. At this pH the active site and the substrate molecule are a perfect fit. The further away from this pH, the less perfect the fit is.

The effect of temperature

As the temperature of a substrate increases, its molecules gain more energy. More collisions occur and the rate of reaction increases. But every enzyme has an optimum temperature at which it works most efficiently. Above the optimum temperature, the enzyme denatures and the reaction stops.

 ...examples of enzymes used in industry

The digestive system

One investigation you can't try

In 1700 Alexis St Martin was shot. The wound did not heal. His doctor was able to see into his stomach and investigate the processes of digestion. Pieces of meat were tied together by a thread and placed into Alexis's stomach. After a few hours the meat had broken down into smaller pieces.

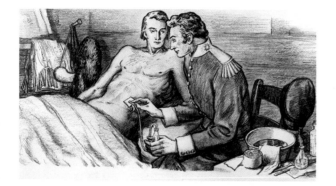

P2

The **digestive system**

Digestion is how your body breaks down large food molecules into smaller ones. It is necessary so that the food molecules become small enough to dissolve into the blood. Your digestive system starts in your mouth and ends at your anus.

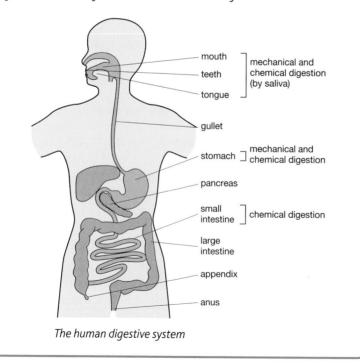

The human digestive system

Know more

Your digestive system is about 9 metres long – nearly the length of a bus. The long passage of food allows time for it all to be digested.

...structure of the digestive system

Know more

Enzymes can have long complicated names but they all end in the same three letters: 'ase'. If you see 'ase' at the end of a word, you know it is an enzyme.

Food enters the mouth where it is chewed up, rolled into a ball by the tongue, and moistened by saliva ready for swallowing.

⬇

Food is swallowed and passes into the gullet which carries the food to the stomach.

⬇

The stomach breaks down food mechanically by muscle contraction and chemically by enzymes. The acid conditions kill bacteria and help the enzymes to work.

⬇

The small intestine digests the food further using different enzymes and absorbs it into the blood.

⬇

In the large intestine water is absorbed to make the waste (faeces) more solid.

⬇

The faeces are then passed out through the anus.

Digestion

Know more

The food in your stomach is not digested by the acid. The acid is needed to kill harmful bacteria and to help the enzymes work.

Mechanical digestion in the mouth

First of all your teeth cut and grind food into smaller pieces. This is an example of **mechanical digestion**.

Chemical digestion in the mouth

Enzymes in your mouth also start to digest the food. This is called **chemical digestion**. It changes the large food molecules into smaller molecules. Different parts of your digestive system have different types of enzymes. Those in your mouth belong to a group of enzymes called **carbohydrases**. These break down carbohydrates into a sugar called **glucose**.

Moving on down

The food in your mouth is mixed with saliva. The tongue then rolls the food into a ball and pushes it to the back of your throat. You then swallow the food and it enters a tube called the **oesophagus**. Muscles in the wall of the oesophagus contract behind the food. This action squeezes the food down. Think of it like squeezing the toothpaste from a tube.

The muscle action is called **peristalsis**. It moves food all the way through your digestive system. It even moves the waste food out at the other end.

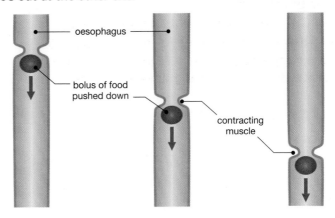

Peristalsis in the oesophagus

Mechanical digestion in the stomach

Mechanical digestion continues in the stomach as the stomach muscles squeeze the food. This action also mixes the food with stomach juice to help it digest. The food stays in the stomach for two hours.

Chemical digestion in the stomach

The stomach juice contains hydrochloric acid and enzymes called **proteases**. These enzymes break down the proteins in the food into smaller molecules called **amino acids**.

P2

🔍 ...mechanical digestion ...chemical digestion

Digestion in the small intestine

The small intestine is where digestion is finished. Carbohydrases and protesases continue chemical digestion, along with another enzyme called **lipase**. Lipase breaks down fats into **fatty acids** and **glycerol**.

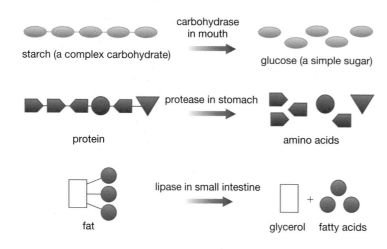

Action of enzymes in the digestive system

Absorption in the small intestine

The food is now made up of small, soluble molecules that will dissolve in the blood. The molecules pass through the thin wall of the small intestine. The small intestine is adapted to **absorb** as much food as possible.

- It has a very thin wall.

- It is covered in structures called **villi** which give the small intestine a very large surface area.

- There are lots of blood capillaries to take the food molecules away.

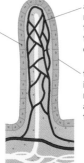

villus wall is only one cell thick; digested food does not have far to diffuse into the blood

a good blood supply means that the digested food is quickly taken away from the villus so more can diffuse across to replace it

the membrane of the villus is permeable to food molecules – this is important as it means they can pass through the membrane

Structure of a villus in the small intestine

 ...absorption in the small intestine

Function of the large intestine

P2

By now most of the food has been broken down and absorbed into the blood. What is left is mainly fibre, a few dead cells and water. Some of the water is absorbed back into the blood. The solid waste, called **faeces**, will be egested through the anus.

What happens to the food?

In the small intestine the food molecules are absorbed into the blood. The blood takes these nutrients around the body where they are used for many processes.

- Glucose is used in respiration to provide energy.

- Amino acids are turned back into proteins to make new cells.

- Fats can be used for energy, or stored under the skin to help keep the body warm.

Research

Find out what might cause a stomach ulcer and how your diet might have to change if you had one.

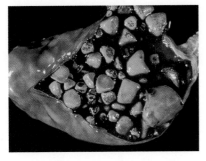

Gallstones inside a gall bladder

Think about

Why would someone with gallstones need to reduce fat in their diet?

Digestive problems

D2

Your digestive system is very important to your health. If it goes wrong you can become ill very quickly. Here are just some problems people have with their digestive system.

Stomach ulcers

The wall of your stomach makes hydrochloric acid. The acid is needed to kill harmful bacteria in your food. However, acid is corrosive so it can damage the stomach wall. To protect itself from acid the stomach produces mucus. Sometimes the stomach makes too much acid and not enough mucus. The acid then damages the wall, forming an ulcer.

Gall stones

To help the digestion of fat, your liver makes a chemical called **bile**. The bile is stored in the gall bladder, ready for use in the small intestine. Sometimes this bile becomes hard and turns into gallstones. These can block the bile duct leading to the small intestine. The blockage is very painful and causes problems with the digestion of fat.

...diseases of digestive system

The respiratory system

Out of breath

Virender has asthma. He has to take medicine from an inhaler to help him breathe. Asthma is a problem for over 5.2 million people in the UK. When they have an attack they become short of breath and start to wheeze. Knowing more about the structure of the respiratory system can help us understand what happens during an asthma attack.

P2

Breathing

All living things need oxygen to survive. Oxygen is taken from the air and used to release energy from food. This is called **respiration**. It produces carbon dioxide, which our bodies need to get rid of.

We use our **respiratory system** to breathe – to take oxygen into the lungs and to expel carbon dioxide from the lungs. This is called gas exchange. The main parts of the respiratory system are shown here.

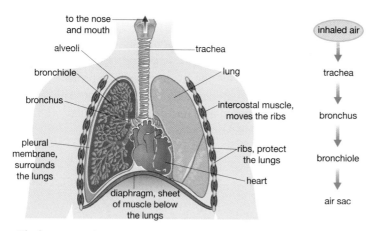

The human respiratory system

When we breathe, air is taken in, or inhaled (**inspiration**) and then forced out, or exhaled (**expiration**). To do this we have to contract and relax muscles in our chest.

The oxygen from the air needs to get into the blood. This happens inside the lungs, in the **alveoli**.

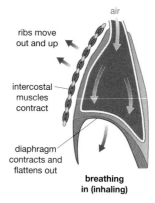

breathing in (inhaling)

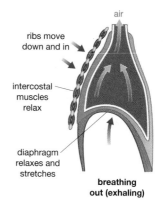

breathing out (exhaling)

Inhaling and exhaling (a side view of the chest)

Know more

Breathing involves the exchange of gases. Respiration is the release of energy from food. Be careful how you use these words.

🔍 ...structure of the lungs

M2

Research

If you could spread out all the alveoli in your lungs they would cover the playing surface of a popular sport. See if you can find out which one. Try typing 'How large is the surface area of your lungs?' into a search engine.

The importance of alveoli

The alveoli are specially adapted for gas exchange.

* There are large numbers of alveoli. This increases the surface area for gas exchange to take place.

* The walls of the alveoli are very thin. So the oxygen can easily **diffuse** into the blood (and carbon dioxide out).

* Each alveoli is covered in tiny blood capillaries to take the oxygen around the body.

Together the lungs and the blood make sure the cells in your body have enough oxygen for respiration.

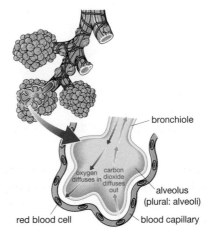

Gas exchange in an alveolus

Think about

Emphysema is a lung condition caused by smoking. The alveoli have a reduced surface area. This means sufferers find it difficult to walk long distances. Can you explain why?

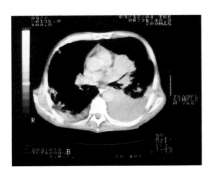

Cancerous tissue (white) in the lungs

D2

Lung problems

Special cells in the lining of the trachea and the bronchioles make sticky mucus. The mucus traps dust particles and some bacteria. The lining is also covered by millions of tiny hair-like structures called **cilia**. The cilia move together in a wave-like motion which carries mucus and any trapped dust upwards into the throat and out of the lungs.

Cigarette smoke stops the cilia from moving properly. Dust collects and irritates the cells. Smokers need to cough to move the mucus upwards.

Smoking is also linked to **lung cancer**. Cells of the alveoli divide in an uncontrolled way. This reduces the working surface area of the alveoli and eventually destroys lung tissue.

 ...diseases of the lungs

Blood

Donating blood

Liam is a blood donor. On each visit to his local blood donation centre Liam can donate 300 cm³ of blood. This is about 5% of the blood in his body. Liam's body can replace this blood in 24 hours. Hospitals need as much donated blood as possible.

P2

What's in blood?

Blood looks like a red liquid. However, if you take a closer look it is very different. Blood is made up of a yellow liquid called **plasma**. The blood looks red because it contains tiny **red blood cells**. These cells transport oxygen around the body. Look even closer and you can find **white blood cells**. These help to defend the body against disease. Blood also contains tiny cell fragments called **platelets**, which help to clot the blood if we cut ourselves.

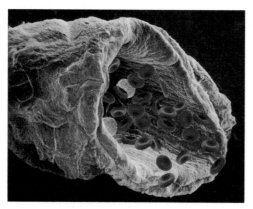

High-magnification image of a blood vessel showing red and white blood cells

Functions of blood

Blood has many important functions. Each part of the blood needs to be adapted to carry out that function.

Know more

*Blood transfusions only work if the blood type of the patient is matched. The different **blood groups**, **A**, **B**, **AB** and **O** were not discovered until 1900. Before that the patient often died of a blood clot caused by mixing blood of the wrong groups.*

...blood structure and function

P2

Red blood cells

Red blood cells are adapted to carry as much oxygen as possible.

- The red colour comes from a chemical called **haemoglobin**. Oxygen joins to haemoglobin inside the cells.

- There is no nucleus in a red blood cell. This leaves more room for the oxygen.

- Red blood cells are disk-shaped with a dent on both sides. This helps them absorb a lot of oxygen.

- They are tiny. They can take the oxygen all over your body.

White blood cells

White blood cells can change shape. They need to do this so that they can wrap around microbes and engulf them. Their flexible shape also helps them to squeeze through capillary walls to reach the microbes.

Plasma

The liquid plasma is also a very important part of the blood. It transports water, hormones, antibodies, food nutrients and waste products around the body.

Know more

Red blood cells are so small that 5 000 000 000 will fit into 1 cm³.

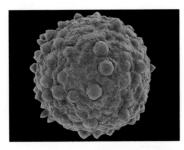

A white blood cell highly magnified

Think about

If we breathe in carbon monoxide, the blood is prevented from carrying oxygen. Which gas do you think is more strongly attached to haemoglobin – oxygen or carbon monoxide?

M2

The importance of haemoglobin

Haemoglobin is a very special chemical in the red blood cells.

- In the lungs it reacts with oxygen to form oxyhaemoglobin.

- Blood containing oxyhaemoglobin travels away from the lungs to other parts of the body.

- When it reaches tissue, the oxyhaemoglobin separates to form haemoglobin and oxygen.

- The oxygen diffuses into the tissue cells and the red blood cells with haemoglobin return to the lungs.

Together the lungs and haemoglobin make sure the cells of the body have the oxygen they need for respiration.

The transport of waste carbon dioxide from body cells to the lungs also involves haemoglobin. However, it is a much more complicated process.

 ...blood groups

The heart and circulation

Your assessment criteria:

P2 Carry out investigations into the structure and functions associated with the circulatory system

Practical

M2 Explain the way the respiratory and circulatory systems interact to maintain cellular and body function

D2 Explain the consequences for the human body when the circulatory system fails

The heart of a pig

In 1968 Dr Ross of the National Heart Hospital in London attempted to transplant a pig heart into a patient. The heart was rejected.

Scientists have now cloned genetically engineered pigs. They hope these pigs will provide new hearts for humans without the threat of rejection.

P2

The circulatory system

The **circulatory system** is a transport system. It takes substances, such as oxygen, around your body in the blood. The blood is transported through blood vessels. Without the circulatory system the body would not be able to function.

Blood vessels

Blood is carried around the body in three different types of blood vessels. Each one has a different function:

- **arteries** transport blood away from the heart

- **veins** transport blood to the heart

- **capillaries** join arteries to veins.

The job of the heart

The heart pumps the blood around the body in the blood vessels.

There are two sides to the heart.

- The right side pumps blood to the lungs.

- The left side pumps blood to the rest of the body.

- The blood leaves the heart at high pressure in arteries.

- The blood returns at low pressure in veins.

> **Know more**
>
> *The body of an adult contains over 60 000 miles of blood vessels!*

> **Know more**
>
> *In 1628 William Harvey became the first person to demonstrate that blood travels in blood vessels.*

 ...heart structure and function

Know more

One adult in the UK dies from heart disease every 3 minutes. It is very important to look after your heart.

There are four parts to the heart, called chambers.

- Two **ventricles** pump blood into arteries.
- Two **atria** receive blood from veins.

There are four main blood vessels linked with the heart.

- The pulmonary artery takes blood from the right ventricle to the lungs.
- The pulmonary vein brings blood to the left atrium from the lungs.
- The aorta takes blood from the left ventricle to the body.
- The vena cava brings blood to the right atrium from the body.

There are also **valves** in the heart. They prevent backflow of blood.

All this information is summarised in the diagram.

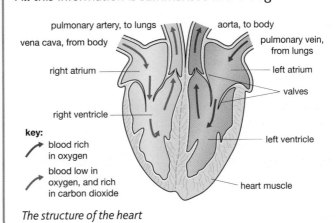

The structure of the heart

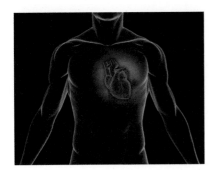

The human heart is not actually on the left side; it just tilts to the left

How the circulatory and respiratory systems work together

M2

The circulatory system in humans is adapted to get as much oxygen to the body cells as possible. To do this it needs to pump blood to the lungs to collect the oxygen and then pump the **oxygenated blood** (blood containing oxyhaemoglobin) to the cells.

Research

Sometimes babies are born with a hole in their heart. Find out why this is a problem.

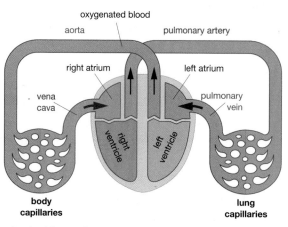

The double circulatory system

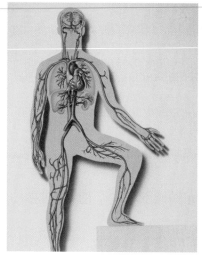

Left and right circulatory system

The left and right sides of the heart act as two different pumps. The right ventricle pumps the blood to the lungs. The left ventricle pumps the oxygenated blood to the cells of the rest of the body. The left ventricle has a much thicker muscular wall than the right ventricle, because it pumps the blood further.

The different types of blood vessel are also adapted for their particular function.

Blood vessel	Adaptation
arteries	Have a thick muscular wall. This helps to withstand high blood pressure as blood leaves the heart.
veins	Have a large lumen (centre hole) to help blood flow at low pressure. Valves stop the blood flowing the wrong way.
capillaries	Have a thin permeable wall. This allows exchange of gases and other substances with body tissue.

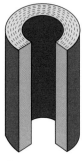

Artery

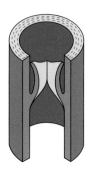

Vein

Capillary

 ...arteries and veins

Circulatory problems

D2

Sometimes the circulatory system does not work as efficiently as it should.

- A fatty diet can lead to a high level of **cholesterol** in the blood. The cholesterol can stick to the inside of artery walls. As it builds up, it forms a **plaque** that affects the flow of blood. Some of this can break away and block the artery completely. If this happens in the coronary artery, oxygen supply to the heart muscle is restricted. This often results in a heart attack.

- Stress and some illnesses can lead to raised blood pressure. This results in a higher risk of a stroke (a blood vessel burst in the brain) and of kidney damage.

- Smoking increases the heart rate, which puts stress on the heart. Smoking also increases cholesterol levels. Both of these factors can lead to heart disease.

- Cigarette smoke contains carbon monoxide. Taking in carbon monoxide reduces the blood's ability to carry oxygen.

- Lack of exercise can lead to circulatory problems. Regular exercise is good for the heart. It can lower blood pressure and improve blood circulation. This helps prevent blood clots that could lead to heart attacks. Exercise also helps reduce the risk of obesity and lowers cholesterol levels in the blood. Both of these factors mean there is less stress on the heart.

Discuss

A hectic, stressed lifestyle with a diet of junk food can reduce life expectancy. Discuss the effect of this lifestyle on the body and why it could result in an early death.

Think about

Smokers often get out of breath much quicker than non-smokers when they exercise. Try to explain why.

Lack of regular exercise could lead to circulatory problems

...high blood pressure

Exercise and the body

Even the fastest get out of breath!

In 2008 Usain Bolt became the fastest 100 m and 200 m runner in the world. To achieve this, Usain's respiratory and circulatory systems had to work perfectly together so that his muscles would get the oxygen they needed. Even after the race these systems would still need to be working harder than normal.

P2

How respiration works

Respiration provides **energy** for our bodies to work. Body cells use oxygen to release energy from **glucose**. The oxygen comes from your lungs to the cells in your blood. The glucose comes from the food you eat and is also transported by the blood.

Respiration makes carbon dioxide. This waste gas is taken to the lungs by the blood and breathed out.

The body uses two different types of respiration.

- Respiration that uses oxygen is called **aerobic respiration**. It is a chemical reaction:

  glucose + oxygen $\longrightarrow$ carbon dioxide + water + energy

- During hard exercise, not enough oxygen gets to the muscle cells. The cells have to carry out **anaerobic respiration**:

  glucose $\longrightarrow$ lactic acid + energy

Less energy is released from anaerobic respiration. It produces **lactic acid**, which collects in muscles and causes pain and fatigue.

Exercise and heart rate

When you are sitting still, you breathe at a steady rate of about 20 breaths per minute. When you exercise, your muscles need more oxygen. They use the oxygen to release energy from food. Your breathing rate needs to increase. And your heart has to pump blood faster around your body, to deliver the oxygen needed.

Know more

Teenagers should exercise at a moderate rate for an hour a day. Moderate exercise means you can still talk while you exercise. If you can sing while you exercise you are not working hard enough.

...why breathing rate increases during exercise

P2

Heart rate is the number of contractions of the heart muscles (heartbeats) every minute. It can easily be measured by measuring your **pulse rate**. Your pulse is the movement of blood in your arteries. If the heart pumps faster, the blood moves faster. This means your pulse rate increases.

Your increased heart rate also helps to get rid of the waste carbon dioxide produced by the muscle cells as quickly as possible.

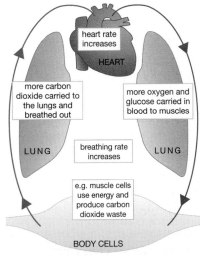

How the lungs and heart work together to get oxygen to the muscle cells

Sprinters running a 100 m race do not usually breathe during the race

Think about

Exercise is not the only factor that can change heart rate. Measure your pulse rate, then drink something with caffeine in it such as coffee or cola. What happens to your pulse rate?

M2

What happens if you can't change the rate enough?

During a sprint race the muscles cannot get oxygen quickly enough. This is because the lungs cannot move air in and out fast enough, and the heart cannot beat fast enough to move enough oxygen to the muscle.

Muscle cells have to use anaerobic respiration to release some energy. An **oxygen debt** builds up.

The energy released during anaerobic respiration is less than that from aerobic respiration, because glucose is only partly broken down.

At the end of the sprint:

* breathing rate is high for a few minutes to replace the oxygen

* heart rate stays high for a few minutes, so blood can carry lactic acid quickly to the liver, where it is broken down.

 ...why heart rate increases during exercise

The kidney

Kidney Camps

Every year, groups of Australian children get together for a special camp. These children have one thing in common — kidney disease. Their kidneys cannot remove waste from their body. They have to spend a lot of time in hospital hooked up to a dialysis machine.

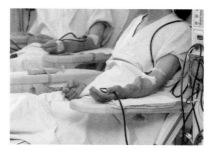

P2

How the kidneys work

Your two kidneys are part of your **renal system**. Kidneys get rid of waste produced by the body. This is called **excretion**.

The rest of your renal system includes the **bladder** and the tubes that connect the bladder to the kidneys.

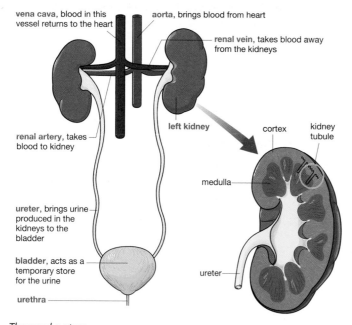

vena cava, blood in this vessel returns to the heart

aorta, brings blood from heart

renal vein, takes blood away from the kidneys

renal artery, takes blood to kidney

left kidney

cortex

kidney tubule

medulla

ureter, brings urine produced in the kidneys to the bladder

bladder, acts as a temporary store for the urine

urethra

ureter

The renal system

The renal artery brings blood containing waste substances to the kidneys. One waste substance is **urea**. It is made in the liver from unwanted amino acids. The kidneys filter the blood, removing the waste. Useful substances such as glucose, some water and some salt are re-absorbed back into the blood. The liquid waste from kidneys is called **urine**.

Know more

Urea is made in the liver and excreted by the kidney.

Urine is made in the kidney.

🔍 ...kidney structure and function

Know more

If you drink large amounts of liquid you will make large amounts of urine.

If you do more exercise or get very hot, more water is lost as sweat so less is lost as urine.

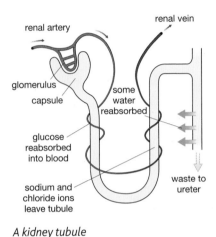

A kidney tubule

Labels:
- renal artery
- renal vein
- glomerulus
- capsule
- some water reabsorbed
- glucose reabsorbed into blood
- sodium and chloride ions leave tubule
- waste to ureter

P2

A balancing act

The function of the kidney is to balance the concentrations of different substances in the blood.

- Water – If the blood contains too much water, the water will enter the body cells. If the cells contain too much water they will burst.

- Salts – These are chemicals such as sodium chloride, the salt you put on your chips. Too much salt in your blood means water will leave the body cells.

- pH – This is a measure of how acidic your blood is. If your blood becomes too acidic, enzymes stop working. The kidney can remove hydrogen ions from your blood, making it less acidic.

Inside the kidney

A kidney contains about half a million **kidney tubules** to filter the blood. Useful and waste materials are filtered from the blood by the **glomerulus**. In the next part of the tubule the blood selectively re-absorbs useful substances such as glucose and some water. Excess water, salts and hydrogen ions will stay in the tubule. The tubules then take the waste substances to the ureter. The ureter takes the waste, in the form of liquid urine, to the bladder where it is stored.

Think about

Ecstasy is a drug that can cause the body to release too much ADH. What would happen to your body cells if the pituitary gland released too much ADH?

Research

Find out about kidney dialysis and why people need it. A good starting point would be:

www.kidneydialysis.org.uk

D2

Controlling water balance

Urine concentration depends on how much water is re-absorbed in the kidney tubules. **Anti-diuretic hormone (ADH)**, made in the pituitary gland in the brain, controls the re-absorption of water. If your body does not release enough ADH you can become dehydrated.

Kidney dialysis

Some people have kidneys that do not work. Toxins in the blood can quickly build up causing them harm. They have to have **kidney dialysis** or a kidney transplant.

...why kidneys fail

The nervous system

Fastest reflexes in the world!

If you touch something very hot, you move your hand away really quickly. But however fast your reactions are, they cannot beat the trap-jaw ant. When set off by touch, the jaw can accelerate from 0 to 143 miles per hour in 0.13 milliseconds. That is 2300 times faster than you can blink your eye.

The trap-jaw ant

P3 The **nervous system**

Your nervous system co-ordinates your body. **Receptors** in your body detect a **stimulus** such as a hot surface. Information is then sent to your spinal cord and brain. From there, information can be sent to an **effector**. Muscles are effectors. They respond to a stimulus by moving your body.

There are two parts to your nervous system:

- the **central nervous system (CNS)**, made up of the brain and spinal cord

- the **peripheral nervous system**, made up of nerves to and from the brain and spinal cord.

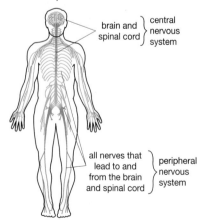

brain and spinal cord ⎫ central nervous system

all nerves that lead to and from the brain and spinal cord ⎫ peripheral nervous system

The nervous system of the human body

Know more

Each of your fingertips has about 100 touch receptors. This is many more than other parts of your body. This is why you use your fingers to feel with.

🔍 ...what the nervous system is made of

P3

Information is carried round the body by mainly electrical impulses carried in nerve cells called **neurons**. The neurons that carry information from receptors to the central nervous system are called **sensory neurons**. The neurons that carry information from the central nervous system to muscles are **motor neurons**.

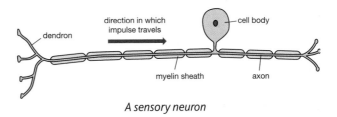

A sensory neuron

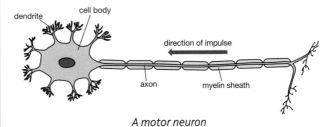

A motor neuron

Reflexes

Reflex actions are those we take without thinking. Reflexes protect us because they are fast and automatic. Examples include the pupil of the eye changing size (getting smaller in bright light), the knee-jerk response (when tapped below the knee) and taking a hand away from a hot object.

A **reflex arc** is the pathway taken by a nerve impulse as it passes from a receptor through the nervous system to the effector. It does not go via the brain.

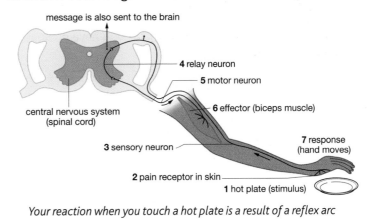

Your reaction when you touch a hot plate is a result of a reflex arc

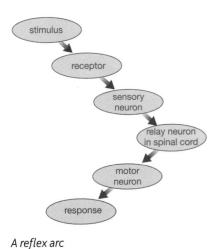

A reflex arc

Know more

Find out how fast your reflexes are at www.happyhub.com/network/reflex/

🔍 ...multiple sclerosis

P3

Synapses

A **synapse** is a gap between two neurons. To carry the information from one neuron to the next, the synapse releases a chemical. When the chemical reaches the other side it causes an electrical impulse to travel along the second neuron.

M3

Voluntary and automatic responses

Your nervous system has two different methods of controlling your body. We call these methods the **somatic nervous system** and the **autonomic nervous system**.

The somatic nervous system is mainly about voluntary actions. It uses your senses to detect changes in the world around you.

The response is controlled by your brain. Muscles attached to your skeleton receive a signal to respond to the changes detected. One example of using the somatic nervous system is catching a ball. Your eyes detect the ball moving towards you. Your brain assesses this information and instructs the muscles of your arm to reach out and catch the ball. The response is not automatic but voluntary: you could decide not to catch the ball.

A voluntary action

The autonomic system controls your internal environment. When receptors detect your body getting too warm, for example, impulses are sent to your skin. The skin responds by releasing sweat. The response is automatic: you have no control over the action.

Reflex actions are part of both the somatic nervous system and the autonomic nervous system.

- Removing your hand from a hot surface is a somatic, not autonomic, reflex because the skeletal muscle is involved, and the brain can override the response. If you pick up a hot plate with your dinner on it, your reflex is to drop it. But if your brain is quick enough it will stop you, because you don't want your dinner all over the floor.

- Sweating when you get warm, and your pupil size decreasing in bright light, are examples of autonomic reflexes. The response is triggered without thought.

The pupil gets smaller in bright light

Think about

What might the response be if your body got too cold?

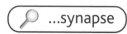
...synapse

Discuss

Explain why it is important not to drive after drinking alcohol.

Think about

Christopher Reeve needed a ventilator to help him breathe. Is it the autonomic or the somatic system that controls breathing?

Research

You are out with your friend and she (or he) has a fall. She may have a spinal injury. Would you know what to do? Find out from: www.redcross.org.uk

Response failure

When someone drinks alcohol their reactions are slowed down. The alcohol affects the synapses so that they cannot pass on the impulse effectively.

The effect of alcohol on the nervous system is temporary. There are other more serious permanent disorders.

In 1995 the actor Christopher Reeve was thrown from his horse and damaged his spinal cord. The damage occurred in his neck and he became paralysed. This meant that nerve impulses could not be sent to his lower body. His could no longer walk and spent the rest of his life in a wheelchair.

Multiple sclerosis (MS)

Multiple sclerosis, or MS, is a disorder of the nervous system that affects around 100 000 people in the UK. It is caused by damage to the **myelin sheath**, which is the protective outer layer around the neurons of the CNS. This damage interferes with impulses travelling along the axon – they can slow down, or not get through at all.

The damage occurs when the body's immune system, which normally helps to fight off infections, attacks myelin.

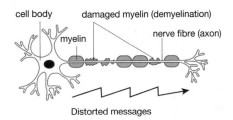

cell body · damaged myelin (demyelination)

myelin · nerve fibre (axon)

Distorted messages

Damage to the myelin sheath causes distorted nerve impulses

Symptoms of MS are different for different people, depending on which part of the CNS has been damaged. Some people get worse as the years go by. Others have lengths of time when they are better and then they become worse again.

Think about

If there is myelin damage on a motor neuron leaving the spine and going to the left arm, what symptoms might there be?

Hormones

Your assessment criteria:

P4 Identify the function of the main endocrine glands

M4 Describe the way hormones co-ordinate body functions

D4 Assess the difference between the way hormones co-ordinate body functions and the way the nervous system co-ordinates body functions

Size matters

Sultan Kosen from Turkey is the world's tallest man. He is 77 cm taller than the average man. His height is due to the medical condition called pituitary gigantism. His body has produced too much growth hormone.

P4

What are hormones?

Hormones are chemicals that control your body in different ways. They are made in special **glands**. Hormones are carried in the blood to a part of the body where they have an effect. The part of the body that a hormone affects is called the **target organ**.

Glands

The glands that produce hormones are part of your **endocrine system**.

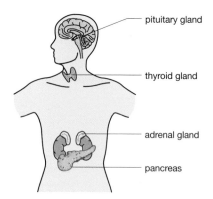

Four of the endocrine glands in the body that release hormones

Know more

The word 'hormone' means to 'excite' which is what hormones do. They cause other things to start happening inside the body, such as causing your heart to beat faster.

🔍 ...the endocrine system

The pituitary gland

The **pituitary gland** is found at the base of the brain. It produces many different hormones. The table shows four of them.

Hormone	What it does
growth hormone	Works with other hormones to control growth
ADH	Controls how much water the kidneys excrete
LH and FSH	Control the production of eggs in females

Important hormones released by the pituitary gland

The pancreas

The **pancreas** makes the hormone **insulin**. This controls the level of glucose in your blood.

When you eat food containing carbohydrates, such as potatoes, bread or cakes, the level of glucose in your blood rises. The pancreas releases insulin. This causes the liver to store some of the glucose.

If there is not enough glucose in your blood, the pancreas releases another hormone called **glucagon**. This causes the liver to release some of the stored glucose.

The adrenal glands

The **adrenal glands** are joined to your kidneys. One of the hormones they make is called **adrenaline**. This is sometimes called the 'flight or fight' hormone, because it is released when you get excited or nervous. Adrenaline helps your body prepare for action by making sure your muscles have enough energy to respond.

Corticosteroids are a group of hormones also made by the adrenal gland. They are important in helping the body cope with stress, such as illness or injury.

The thyroid gland

The **thyroid gland** makes a hormone called **thyroxin**. Thyroxin controls your metabolic rate, which is the rate at which you use energy in your body. If your thyroid releases too much thyroxin, your body quickly uses up energy. People with an overactive thyroid find it difficult to sleep and tend to lose weight.

Know more

Some people cannot control their blood glucose levels. They have **diabetes**. There are 2.3 million people in the UK who have been diagnosed with diabetes. This number is increasing each year.
The increase is probably partly due to a diet too rich in sugars.

Know more

'Derbyshire neck', also called goitre, is a swelling of the thyroid gland. In the 18th century it was very common in the Derbyshire Peak District. The reason is now known to be a lack of iodine in the diet. Iodine is now added to drinking water to prevent goitre.

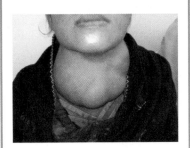

Swelling of the thyroid gland

🔍 ...the effect of adrenaline on the body

How do hormones work?

Hormones are involved in **homeostasis**. Homeostasis is the way the body keeps a constant internal environment. The body's use of the hormone ADH to control a constant level of water in the blood (see page 00) is one example.

Homeostasis involves **feedback mechanisms**. If the internal environment changes, the mechanism is 'switched on' – for example, a hormone is released. When conditions return to normal, the mechanism is 'switched off' – no more hormone is released.

Insulin and blood sugar level

The use of hormones to control blood glucose levels is an example of a feedback mechanism.

- After a meal containing carbohydrates, the digestive system breaks the large carbohydrate molecules down into glucose molecules. Glucose is carried in the blood to body cells.
- Cells in the pancreas detect the higher levels of glucose. They secrete insulin, which travels in the blood to the liver.
- Receptors on the membranes of the liver cells detect the insulin. The insulin causes the liver cells to take the glucose out of the blood. The glucose is changed into a chemical called glycogen and stored.
- When the level of glucose in the blood returns to normal, the pancreas stops secreting insulin. The mechanism has been switched off.

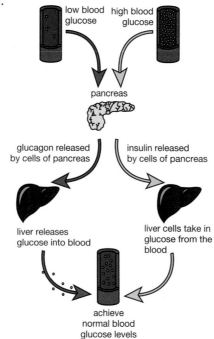

The mechanism for control of glucose level in the blood

Think about

Why is a thermostat in a boiler an example of a negative feedback mechanism?

Research

Find out the symptoms of someone who suffers from diabetes.

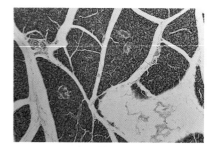

Insulin is made in the cells of the pancreas

Think about

Look at the diagram on the left. Describe what would happen if the glucose level in the blood got too low.

...diabetes

M4

Adrenaline action

Adrenaline prepares your body for action in many different ways. These are just some of the changes it causes:

- More glucose is released into your body, to give you energy.

- Your heart rate is increased, so your muscles get more oxygen and glucose.

- The airways in your lungs widen, so more air reaches the alveoli.

- Blood vessels taking blood to your muscles widen, and those in your digestive system narrow. This diverts the blood to where it is needed most.

Thyroxin action

Thyroxin is needed to co-ordinate many processes in the body. It works together with other hormones and with the nervous system. When you get too cold, thyroxin increases heart rate and respiration rate so that energy from food is released as heat. Thyroxin also helps you grow, by controlling the growth of your bones and the production of proteins.

Think about

The effects of the endocrine system are usually slow and long-lasting. Which hormone produces a quick, short-lived response?

Research

Your nervous system can help out your endocrine system when you need to run from danger. Find out how by searching the web for 'fight or flight response nervous system'.

Hormones and nerves compared

D4

Both your nervous system and your endocrine system control your body. Differences in the way they work are shown in the table.

Nervous system	Endocrine system
Electrical impulses carry the information around the body.	Chemicals carry the information around the body.
The information is carried along nerves (neurons).	The information is carried in the blood.
The effects are quick and normally only last for a short time.	The effects are usually slow and longer-lasting.
Often involves reflex reactions.	Controls long-term processes such as growth, development and metabolism (production and use of energy).

Reproduction

Your assessment criteria:

P5 Identify the structure and functions of the male and female human reproductive systems

M5 Explain the process of hormonal control of the female reproductive cycles

D5 Explain the way conception is controlled using replacement hormones

The male pill?

Men could soon have their own version of the contraceptive pill. They could be given a monthly injection of testosterone that would prevent sperm production. Once injections are stopped, normal sperm production will start again. When the injections were trialled in China, only one man in 100 fathered a child.

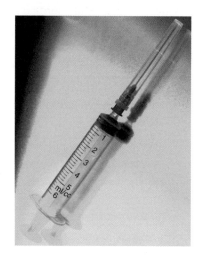

P5 — The human reproductive system

Female humans produce **eggs**. The female reproductive system has:

- **ovaries** that make the eggs (or ova)
- **oviducts** (tubes) to carry the eggs to the right place for fertilisation
- a **uterus** where the baby develops
- a **vagina** through which the baby is born.

Male humans produce **sperm**. The male reproductive system has:

- **testes** to make the sperm
- a **scrotum** to hold the testes
- **sperm ducts** to carry the sperm
- a **penis** which ejects the sperm into the female's vagina.

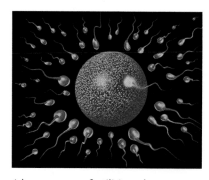

A human sperm fertilising a human egg

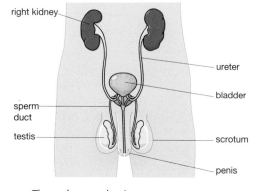

The male reproductive system

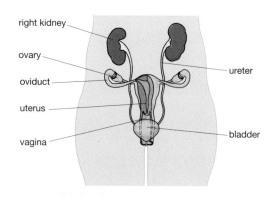

The female reproductive system

...reproductive organs of males and females

Know more

*133 million babies are born in the world each year. But some couples are unable to have children. One or both of them is **infertile**.*

Fertilisation

Human eggs are microscopic. Sperm are even smaller than the eggs. The eggs and sperm are called **gametes**.

Inside the female's body, a sperm can meet and fuse with an egg. This is called **fertilisation**. The genetic information of both gametes combines to produce a new individual. The **foetus** (young baby) begins to develop.

The developing foetus

The foetus develops inside the mother's uterus during pregnancy. A human baby takes 38 weeks to fully develop. Throughout this time the mother provides the foetus with food and oxygen via the **placenta** and **umbilical cord**. These also allow carbon dioxide and other wastes to pass from the foetus's blood to the mother's blood. The fluid in the **amniotic sac** protects the foetus from knocks.

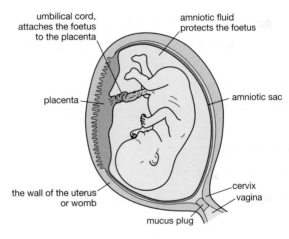

umbilical cord, attaches the foetus to the placenta

amniotic fluid protects the foetus

placenta

amniotic sac

the wall of the uterus or womb

cervix

vagina

mucus plug

Foetus in the womb

Birth

When the baby is ready to be born, the mother goes into **labour**. The muscles of her uterus start to contract. The amniotic sac will break, releasing the fluid. The baby is now ready to be delivered, normally head-first.

The baby's head is pushed towards the **cervix** by the contracting muscles. This can continue for several hours. Once the baby is pushed out, the umbilical cord can be cut. A few minutes later the **afterbirth** (placenta) passes out.

The newborn baby

🔍 ...human embryo development

The female menstrual cycle

When a female reaches puberty her ovaries start releasing eggs, one at a time in a regular cycle lasting about 28 days. This is called the **menstrual cycle**. The release of an egg from the ovary is called **ovulation**. At ovulation the uterus lining starts to become thicker, with more blood vessels. This will help a fertilised egg to embed in the lining. The egg may not be fertilised or embed in the uterus lining. The uterus lining then breaks down, releasing the broken down cells in a flow of blood from the vagina. This is called **menstruation**.

Some couples need fertility treatment before they can have a baby

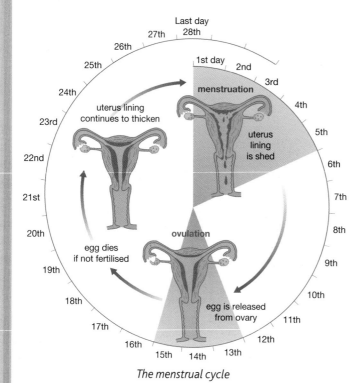

The menstrual cycle

Hormonal control of the menstrual cycle

The menstrual cycle is controlled by hormones. The cycle is triggered by receptors in the hypothalamus in the brain. These cause the pituitary gland to produce two hormones:

- **follicle-stimulating hormone (FSH)**, which stimulates a **follicle** in an ovary to start developing into a mature egg
- **luteinising hormone (LH)**, which controls the release of an egg.

As the follicle in the ovary develops, it releases varying amounts of two more hormones, **oestrogen** and **progesterone**. High levels of these hormones cause the uterus lining to thicken. When their levels start to fall, menstruation starts.

Discuss

Some women do not release enough hormones to enable them to get pregnant. Which hormone could be given to help them release more eggs? What could be the disadvantage of this treatment?

🔍 ...fertility treatments and hormones

Think about

If no fertilisation takes place, oestrogen and progesterone levels in the body fall. Why is it important that this happens?

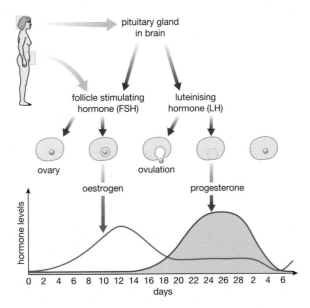

Changing levels of oestrogen and progesterone during the menstrual cycle

M5

If an egg is fertilised:

- the progesterone level remains high, so the uterus lining does not break down

- the oestrogen level remains high, preventing FSH from being released

- no more eggs develop or are released for the duration of pregnancy.

Research

There are different types of contraceptive pill, containing different levels of hormones. Find out about the hormone in the 'mini-pill'. How is it different from the more common 'combined' pill?

D5

Preventing pregnancy

Some women want to be able to have sexual intercourse without running the risk of getting pregnant.

A woman can be prescribed contraceptive pills. The pills contain progesterone and oestrogen. Oestrogen stops FSH being produced, so her eggs do not mature. This means there is never an egg to be fertilised.

Instead of taking pills daily, contraceptive hormones can be given in other ways.

- A patch containing progesterone and oestrogen can be worn on the skin for three out of four weeks. The hormones are absorbed through the skin and enter the blood.

- A hormone injection can be given, which lasts for three months. The disadvantage is that it can affect the normal menstrual cycle, and may affect fertility for a time after a woman stops having the injections.

...contraception

Assessment and Grading criteria

To achieve a pass grade, my portfolio of evidence must show that I can:

Assessment Criteria	Description	✓
P1	Outline the role of enzymes as catalysts	
P2	Carry out investigations into the structure and functions associated with the digestive, respiratory, circulatory and renal systems	
P3	Identify the components of a simple reflex arc	
P4	Identify the function of the main endocrine glands	
P5	Identify the structure and functions of the male and female human reproductive system.	

Studying the human body is an important part of science

To achieve a merit grade, my portfolio of evidence must show that I can:

Assessment Criteria	Description	✓
M1	Explain the factors affecting the functions of enzymes	
M2	Explain the way the respiratory and circulatory systems interact to maintain cellular and body function	
M3	Describe the difference between the somatic and autonomic nervous system	
M4	Describe the way hormones coordinate body functions	
M5	Explain the process of hormonal control of the female reproductive cycles.	

To achieve a distinction grade, my portfolio of evidence must show that I can:

Assessment Criteria	Description	✓
D1	Analyse data to identify the optimal conditions of at least two parameters for the function of an enzyme	
D2	Explain the consequences for the human body when one of these systems fails	
D3	Give possible causes of failure of the nervous system and explain the consequences	
D4	Assess the difference between the way hormones coordinate body functions and the way the nervous system coordinates body functions	
D5	Explain the way conception is controlled using replacement hormones.	

LO

Be able to use appropriate scientific techniques to analyse evidence which has been collected from the scene-of-crime

- Fingerprints have been found that date back thousands of years to the time of the ancient Egyptians, but it was the Chinese who first used fingerprints to verify official documents 1300 years ago

- In 1910, Edmond Locard set up what is thought to be the first police crime laboratory in Lyons in France; he used the phrase, 'every contact leaves a trace'

- A major role of the forensic scientist is to detect and identify recreational drugs; in London, a study showed that 99% of all bank notes were contaminated with cocaine!

- DNA fingerprinting was first used in the 1980s when it was used as evidence to convict murderer Colin Pitchfork

- DNA fingerprinting has revolutionised forensic science – we can now obtain a full DNA profile from a few body cells or flakes of dandruff

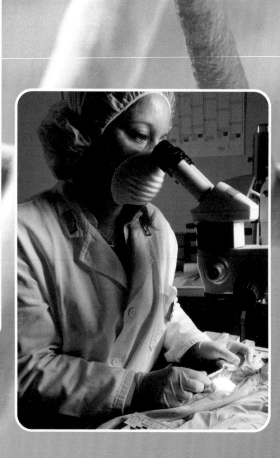

LO

Be able to investigate a scene-of-crime

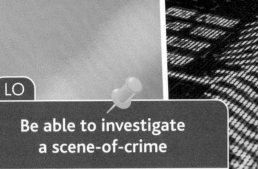

- The Scenes of Crimes Officers (SOCOs) that seal off, record and collect evidence from crime scenes are civilians employed by police forces and not police officers

- SOCOs are called to crime scenes at any time of the day or night, and may have to remain there for days or even weeks so that every piece of evidence is collected

- Many SOCOs now use photographic and surveying techniques to produce virtual reconstructions of crime scenes that help with their recording and investigation of the scene

LO

Understand the relationship of forensic science to the law, including the criminal justice system

- The most serious crimes, such as murder, rape, death by dangerous driving are brought to trial in the Crown Court, where individuals are tried by jury

- When a judge passes sentence, he or she must not only think about punishing the offender, but also about rehabilitating him or her back to a normal life in society

Assessing the crime scene

Your assessment criteria:

P1 Carry out an investigation to collect evidence from a crime scene

> Practical

P2 Demonstrate the most appropriate methods to record and preserve evidence from the crime scene

M1 Describe the processing of a crime scene, explaining how the techniques used obtained valid forensic evidence

D1 Evaluate the processing of a crime scene, interpreting how the valid evidence collected could be used in a criminal investigation

A body is found

A member of the public has found a body in a house. She reports to the police that she thinks the person has been murdered. The crime desk sends a police officer to the incident scene.

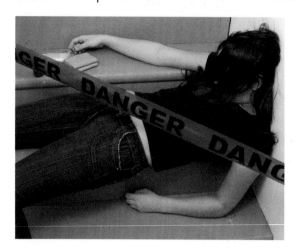

P2 Assessing the scene

On arrival, the police officer's first job is to carry out an **initial assessment** of the scene. If he (or she) is at first unsure whether or not a crime has taken place, it's best to assume that it has. Valuable time and evidence in the investigation could be lost otherwise.

First he must deal with anyone at the scene needing medical help. Any other people present at the scene must be kept apart, as they may be eyewitnesses or suspects. Witnesses at crime scenes are not allowed to talk to each other. A person's perception of what happened can get distorted during conversation.

The police officer must cordon off the area with police tape and make sure that no-one without the appropriate authority can enter. It's also vital that no evidence is removed from – or added to – the scene. He then calls for a **Scene of Crime Officer (SOCO).**

Preserving the scene

When the SOCO arrives at the crime scene, he (or she) puts on a full protective body suit, gloves, a mask, and plastic overshoes. Without this, the SOCO's skin cells, hair, fibres, fingerprints or shoeprints could be added to the crime scene.

Know more

The police officer must enter the crime scene in order to carry out the initial assessment. His health and safety, and that of other people present, is of major importance. He must assess the risks of harm to himself or others present, before entering the scene.

Health and Safety

How safe is it for the investigating team to enter the house? There may be sharp objects or dangerous chemicals, such as cyanide. In cases of terrorism, biohazards such as anthrax, bacteria or radioactive materials used to make a 'dirty bomb' could be present.

 ...police risk assessment

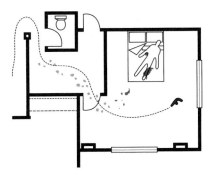

A SOCO's plan of the crime scene gives investigators a permanent record

Research

The value of digital photography for evidence in court has been questioned. Research the advantages and disadvantages of digital photography for recording crime scenes.

Think about

Explain in detail how the protective clothing worn by the SOCO prevents contamination of the scene. Why does the SOCO put this on only when he or she arrives at the scene?

The SOCO in protective clothing

P2

In the case of a death, the SOCO sets up a single entrance to the crime scene and path to the body that everyone must use. This is called the **common approach path**. It's important that the investigating team disturbs the scene as little as possible when entering or leaving.

Recording the scene

The SOCO must produce a permanent record of the crime scene, using detailed written notes, sketches, photographs and videos.

It is essential that the original position of items at the scene is recorded. Some biological and chemical evidence may quickly deteriorate. Other evidence may be very fragile, and might be destroyed as the SOCO tries to recover it.

Other evidence from the scene of crime will be sent to the forensic lab for analysis. The SOCO is aware that objects that don't seem important at the time can turn out to be so later on.

Photographing the scene

When photographing a crime scene, the SOCO follows four rules:

1 Photograph everything.

2 Photograph each item at the scene before doing anything to it.

3 Add a scale and photograph the item again.

4 After collecting trace evidence from the item, or removing it for analysis, photograph the same part of the crime scene again.

Avoiding contamination and loss of evidence

M1

If the correct procedures are not followed, the crime scene could become contaminated by investigators. But this is rare. As well as a SOCO's protective clothing, transparent stepping plates can be used along the common approach path, as shown in the photograph on the left.

Techniques used to record crime scenes

D1

Many police forces are now able to produce **virtual reconstructions** of crime scenes. For investigators, these are better than just photographs. They don't include unnecessary detail, and the investigator gets a better sense of perspective of the crime scene.

🔍 ...crime scene virtual reconstruction

Collecting evidence

Your assessment criteria:

P1 Carry out an investigation to collect evidence from a crime scene

> Practical

P2 Demonstrate the most appropriate methods to record and preserve evidence from the crime scene

M1 Describe the processing of a crime scene, explaining how the techniques used obtained valid forensic evidence

D1 Evaluate the processing of a crime scene, interpreting how the valid evidence collected could be used in a criminal investigation

Every contact leaves a trace

The Scene of Crime Officer (SOCO) has recorded a scene of possible murder. Now the team begins the search for evidence.

When a crime is committed outdoors a canopy is put over to protect and secure the scene

P2

The search for evidence

Any evidence at the crime scene may turn out to be important at some stage in the investigation, so it's important that the team's search is thorough and systematic. SOCOs often cover the whole area with a grid or spiral pattern, so there is less chance of missing things.

- Some of the evidence, such as a cigarette butt, may be immediately obvious to the SOCO.

- Some of the evidence, such as fibres, may be present in very small amounts. This is called **trace evidence**.

- Other evidence, such as fingerprints, may be invisible to the naked eye, and special techniques are needed to reveal it.

- Some evidence may have been damaged, for example burnt. Special procedures are then needed.

Collecting the evidence

The SOCO uses a range of techniques to collect evidence, designed to collect different types of evidence effectively and efficiently. Forceps, sticky tape and vacuuming can be used to collect trace evidence such as hair and fibres. Care must always be taken to prevent any contamination of the evidence.

Evidence in a burnt-out car needs special treatment

A swab similar to a cotton bud can be used to collect blood or other body fluids

 ...trace evidence

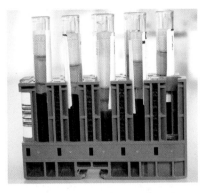

Samples for DNA analysis are kept in special freezers

Storing the evidence

P2

Each item of evidence must be packaged separately, labelled and sealed before it is stored. Small items, such as hairs, fibres, glass fragments and paint, are put into plastic bags or bottles and sealed. Clothing and shoes are put into paper sacks.

Evidence must be stored in secure facilities. Most types of evidence are kept in cool, dry rooms. Biological samples are refrigerated or frozen to prevent their decay.

Think about

Explain why a SOCO collecting evidence from the crime scene doesn't also collect evidence from a suspect that is apprehended at the time.

Avoiding cross-contamination

M1

Trace evidence such as fibres and hair are easily transferred during a crime. Similarly, it could easily be transferred during the investigation. It's vital that there's no **cross-contamination** of one piece of evidence with another.

The chain of continuity

As it's collected, examined and analysed, pieces of evidence are passed from one investigator to another. This is called the **chain of continuity**.

The bag or bottle containing the evidence must be clearly labelled with details of each investigator handling the evidence. Information is added to the label at every stage of the investigation.

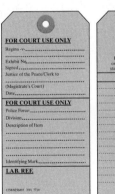

An exhibit label

Discuss

Why is evidence kept for many years after a criminal investigation?

Using the evidence in a criminal investigation

D1

SOCOs must always use standard methods to process evidence. Only then will it provide valid information that can be used, or be **admissible**, in court.

If fingerprint and DNA evidence are absent, incomplete or damaged, other types of evidence may be very important in solving the crime. The way in which the SOCO team searches for, collects, packages and stores such evidence is important in preserving it. Badly preserved evidence may not provide useful information for the investigation and may not be admissible in court.

🔍 ...cross-contamination of evidence

Planning forensic analysis

Your assessment criteria:

P3 Produce a simple plan to analyse biological, chemical and physical evidence from the crime scene

M2 Produce a detailed plan to analyse biological, physical and chemical evidence from the crime scene

The work of a forensic scientist

Anita is a forensic scientist. After a possible murder in a local school, she is sent samples from the crime scene for analysis.

P3

Planning a forensic science investigation

Anita has to plan her **forensic investigation** of the evidence carefully to ensure that valid information can be obtained from it. She tells us more about how she goes about her planning.

My plan

The type of evidence

The standard procedure I will use

The equipment and chemicals that will be required

The observations and/or measurements that I need to make

How I will record the results of the investigation and display the results I've obtained

The time that I'll need to carry out the investigation

'I often have a team to carry out some parts of the practical work. It's essential to get organised so that every member of the team knows what they're doing. If I have no expert knowledge of a particular type of evidence, for instance insects found on a body, I will pass it on to an expert.'

Avoiding contamination

'While carrying out my investigation, I have to make sure that I don't allow evidence to become contaminated. It would be very easy to introduce my own DNA or add fingerprints to an item.

'Any of the evidence that I haven't used for my testing is either left in the sealed evidence bag or returned to it. For evidence I have handled, I sign, date and reseal the bag.'

Know more

Anita must use accepted procedures to carry out all her analyses. Her laboratory is accredited to carry out DNA fingerprinting as part of forensic investigations. Without this accreditation, her findings would not be accepted in court.

Items are packaged separately in evidence bags and sealed so that they don't contaminate each other

 ...crime investigation plan

Know more

The conclusions that the forensic scientist draws are those found out from testing each piece of evidence. They are not about the guilt of a person.

Think about

Anita points out that 'Investigations showing evidence of poor planning will be picked up in court. A defence lawyer might then try to imply that all of a forensic scientist's work is careless.'

What effect would this have on the case?

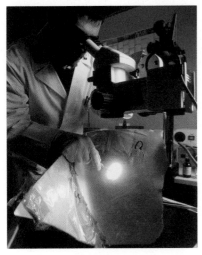

Taking a representative paint sample from a car chassis

Research

Find out about controls or standards used in chromatography to identify unknowns.

Keeping records of the investigation

P3

'For some items of evidence that I've investigated, I decide whether or not I should keep a permanent record. This could be a microscope slide with a hair sample, or cast of a tool mark. This 'new' evidence also needs to be packaged, labelled and stored appropriately.

'I have to record all my findings very carefully. I have blank results tables ready to complete. Later, I will transfer this information to my report that will be used in court. Sometimes I have to make careful calculations. I also have to think about the best way of displaying the data.'

Planning in detail

M2

It's likely that a forensic scientist will receive different types of evidence – biological, chemical and physical. Different techniques, ranging from using a microscope to chemical analysis, will be needed. But for each type of evidence, she must use standard procedures that will be accepted by the courts. The most appropriate method of analysis must be chosen, and this must provide scientifically valid information.

When analysing **chemical evidence**:

- first the forensic scientist undertakes a **qualitative analysis** to tell her what's in a sample

- then it might be important to find out how much is in the sample, so she would do a **quantitative analysis**.

Several techniques may be used on one piece of evidence. For example, the analysis of a fibre may be started by examining it with a microscope, but then an instrumental technique is used to confirm the chemical it's made from. For each test, a **representative sample** of the material needs to be taken. Repeat tests will be carried out if necessary.

Often there's only trace evidence, so the forensic scientist has to use all of the available material for analysis. She must take only very small samples of it if she needs to carry out several tests.

Experimental controls

It may be necessary or helpful to include experimental controls. For example, if a joyrider crashes a car, **control samples** of material from the car, such as glass from a smashed windscreen, can be compared with samples found on the suspect's clothes.

Analysing the evidence

Results, conclusions and risks

Anita, a forensic scientist, has to make very detailed records of her analytical tests and her conclusions.

Your assessment criteria:

M2 Produce a detailed plan to analyse biological, physical and chemical evidence from the crime scene

D2 Assess the potential risks associated with analysing biological, physical and chemical evidence from a crime scene

M2

Detailed records of the analyses

Anita tells us about the records she makes.

'I must record my results carefully. These might be observations or measurements. Even if they're just observations, it's much better if I can also include measurements or scales. I may have to work out the mean of a set of data, such as concentration values.'

The data that a forensic scientist collects must be accurate and precise.

- **Accuracy** is how close a measurement or reading is to its true value.

- **Precision** is the closeness of a series of results to one another.

The accuracy and precision of data collected will depend on the measuring equipment and the chosen techniques.

Processing results and drawing conclusions

Sometimes it may be necessary for forensic scientists to use **statistics** in their analyses. Their work may involve demonstrating whether or not different samples came from the same source, for example from the suspect. The innocence of someone may depend on the samples being *different*.
But apparent differences in sets of data can occur by chance. By carrying out a statistical analysis, it's possible to work out the probability of any differences having occurred by chance.

Conclusions that a forensic scientist draws must be based entirely on the evidence. Sometimes there are limits to the conclusions that can be drawn. This can be because of the quality of the evidence, or because of limitations of the practical techniques.

Forensic scientists use vernier callipers to measure objects to a high degree of accuracy

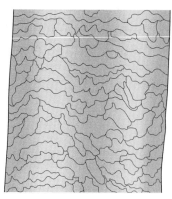

This micrograph of hair needs a scale to indicate its size

Think about

Why must the limitations of the analysis be made clear in the forensic scientist's report that will be presented in court?

...analysing forensic evidence

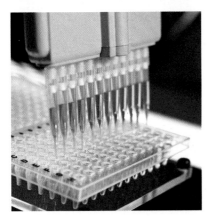

Automated pipettes are used to measure out volumes of liquids accurately

The risks when analysing evidence

D2

A **hazard** is any chemical substance, biological material, radiation, piece of equipment or other item that is likely to cause harm. The **risk** is the likelihood of that hazard causing harm.

Anita talks about the risks in her work.

'During all of the practical work that I carry out, I must minimise the risks to myself, my team and others involved in the investigation. Forensic evidence may also be handled by the magistrate, members of the jury, legal teams and the judge in court. So these people need also to be made aware of any risks in handling the material.

*'For all the work I'm doing, and for evidence that people have to handle, I assess the risk. To do this, I produce a **Risk Assessment.**' An example is shown below.*

Name of hazard, e.g. type of body fluid or living organism	Type of hazard and risk associated with hazard	Steps taken to minimise risk	Emergency procedure
I first list all the potential hazards, even if they're low. I check databases and datasheets listing the hazards of using various materials.	What type of hazard is it, e.g., a corrosive chemical or biohazard? I then have to assess the risk. It's important to think about how I'm using the hazard. And also to think about possible contact of the hazard with a person's body, for instance their skin or eyes. It's really important to take into account the concentrations of substances I'm using. At low concentrations, the risk might be low. But at a high concentration, the chemical might be hazardous.	I must include ways of minimising risk. It's probably better to use a safer, alternative chemical, but this is often not possible. I provide information on the use of personal protective equipment (goggles, lab coats and gloves) and control measures such as fume cupboards.	I must know about emergency procedures in case things were to go wrong.

🔍 ...crime risk assessment

Analysing hair samples

Your assessment criteria:

P4 Carry out experiments to analyse biological, chemical and physical evidence from the crime scene

> Practical

M3 Describe the patterns found from the evidence and make connections

D3 Explain the patterns found from the evidence and make connections.

Samples from a body

Several strands of hair have been found on the clothing and body of a woman who has been killed. There seems to be more than one type of hair, and it is not the same colour or texture as the woman's. The SOCO has taken samples which have been sent to the forensic scientist for analysis.

P4

Examining hair

The forensic scientist examines the hair samples with an ordinary light microscope. From this examination, she will be able to conclude whether the sample came from a human or some other animal. (Hair is a characteristic of all mammals.) If it is human hair, important information can be obtained about the person it came from.

The forensic scientist mounts the hair on a microscope slide in a chemical called glycerol. This allows a clear, detailed view of the layers of the hair.

Hair from a crime scene often needs to be compared with a sample collected from a suspect, or from the suspect's clothing. If the forensic scientist uses a **comparison microscope**, images from the two hairs can be viewed side-by-side.

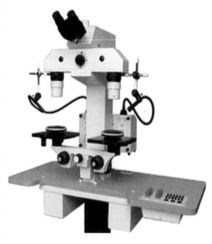

The comparison microscope is basically two microscopes combined into one so that two images can be viewed at the same time

The structure of hair

Hairs grow out of structures in the skin called hair follicles. A hair is attached to the hair follicle by its root.

The parts of the hair that can be seen easily with a microscope are the cuticle, the cortex and the medulla.

A forensic scientist collects hair by tape-lifting it

Know more

*Most of the hair on your head is actively growing – it's in the **anagen phase**. Hairs that are in the **telogen phase** have stopped growing, and we each shed around 100 of these hairs every day. These are easily transferred by physical contact.*

...hair micrograph

Research

Do some internet research to find out how the structure of the hair varies:

- *when taken from different parts of the human body*
- *in different racial groups*
- *between humans and other mammals such as cats or dogs.*

The **medulla** is an air-filled channel that runs down the centre of the hair. This channel is often broken into sections, and has a different structure in other mammals

The **cortex** of the hair is made up of cells containing keratin and the pigment granules that give the hair its colour

The **cuticle** is the outer, protective layer of the hair. It consists of overlapping scales of a protein called keratin

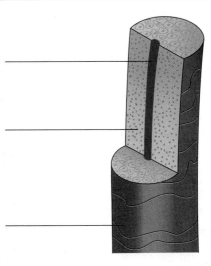

Human hair structure

Think about

An investigating officer removes 30 to 50 hairs from a suspect for comparison with those collected at the crime scene. Why do you think the forensic scientist needs so much hair to examine?

Identification and comparison of hair

From the hair **diameter** and structure, particularly the structure of the medulla, the forensic scientist can often tell conclusively the species of animal that the hair is from.

Analysis of human hair gives information about:

- the part of the body the hair is from
- the colour, length and thickness of the hair
- the ethnicity of the person
- if the hair is natural, and if and when it has been dyed or bleached
- if the hair has been cut recently
- if the person is healthy or not, has a fungal infection or has nits.

Research

Analysis of hair samples can also reveal if a person has been taking drugs. Find out how forensic scientists carry out these analyses. How can this information help crime investigations?

What other information does hair provide?

Hair evidence can tell the forensic scientist if the incident was violent. Hair that has been pulled out in a struggle will mostly still be in its growing phase. It will have roots of a different shape to those of hair that's fallen out.

Hair samples that were in their growing phase may have follicle tissue attached. Then forensic scientists can produce a **DNA fingerprint** from the follicle cells. This can *prove*, beyond any reasonable doubt, that a particular person is associated with a crime scene.

🔍 ...hair structure

Collecting fingerprints

Your assessment criteria:

P2 Demonstrate the most appropriate methods to record and preserve evidence from the crime scene

P4 Carry out experiments to analyse biological, chemical and physical evidence from the crime scene

Practical

A vicious attack in a pub

A fight breaks out in a crowded pub and an innocent person is attacked with a beer glass. In the crowd, it isn't clear who the attacker was, and the people who were involved have run off before the police arrive.

A fingerprint expert arrives at the pub and looks for fingerprints on the beer glass.

P4 What are fingerprints?

Fingerprints are the patterns formed by ridges in the skin on our fingers and thumbs.

- If we accidentally touch a soft surface, such as wet paint or soft putty, impressions of our fingerprints are made. Fingerprint experts call these **plastic fingerprints**.

- Sometimes we transfer a liquid or soft substance that we've come into contact with – such as paint, ink or chocolate – onto another surface. These are called **visible fingerprints**.

- But most fingerprints are formed from the thin layer of sweat and oil that's on our skin. We leave these on *everything* we touch. These are usually invisible. They are called **latent fingerprints**.

Know more

Of the many millions of fingerprints examined over the last 100 years, no two people have been found with the same fingerprints.

P2 Recovering fingerprints for investigation

Plastic or visible fingerprints at a crime scene can be easily recorded by taking photographs. It's much more difficult to find latent fingerprints, which could be anywhere at the scene. SOCOs or fingerprint experts use ultraviolet light or laser light to illuminate objects and show up the fingerprints.

Fingerprints fluoresce under ultraviolet light

...fingerprinting UV light

Know more

Sometimes fingerprint experts have large areas to search. One of the best techniques is then Superglue® fuming. This can reveal fingerprints in a stolen car in ten minutes – powdering would take hours.

Ninhydrin has been used to show up prints that were left 30 or more years ago

Know more

An even better technique now used is to spray a chemical called DFO on the surface and then view the fingerprints with a laser. DFO can detect minute amounts of amino acids in fingerprints – it's ten times more sensitive than ninhydrin.

P2

The simplest and most widely used technique to collect latent fingerprints on *hard surfaces* such as tables, door frames and hard objects is **powdering**. The fingerprint expert applies a special powder using a soft brush. The powder sticks to the sweat and oil of the fingerprint. When the excess powder is carefully removed, the fingerprint will be revealed.

The powder is selected according to the surface to be investigated. Aluminium powder is the most commonly used. Carbon black powder is good for use on light-coloured surfaces.

Magnetic powders are particularly useful for collecting prints from human skin or leather. The powder is applied with a special applicator, and an inbuilt magnet in the applicator then removes the excess powder.

Important fingerprint evidence is often left on porous surfaces such as paper or plasterwork. A fingerprint will soak into the material instead of sitting on its surface, and powdering will not show it up. One technique used in these cases involves spraying the surface with a chemical called *ninhydrin*. This reacts with amino acids in sweat, and fingerprints appear in shades of purple and blue.

Applying magnetic fingerprinting powder

The importance of fingerprint evidence

As far as we know, everyone's fingerprints are unique. Not even identical twins have the same fingerprints. And, except for their size, fingerprints remain unchanged during a person's life. They are also one of the last features to decompose after death, so they can be used in post-mortem identifications.

...micrograph of fingerprint

Analysing fingerprints

Your assessment criteria:

P4 Carry out experiments to analyse biological, chemical and physical evidence from the crime scene

> Practical

M3 Describe the patterns found from the evidence and make connections

D3 Explain the patterns found from the evidence and make connections.

The search is on

A man has been seriously injured with a beer glass in a pub. A fingerprint expert called to the scene has collected a set of fingerprints from the glass. The search is now on for the attacker.

P4 Comparing fingerprints

An analysis of fingerprints is only of use if the fingerprint expert can make a comparison. Fingerprints from the scene will be compared with those of a suspect or those of known criminals from a database.

When a suspect is apprehended by the police, a set of their fingerprints is recorded. For 100 years this was done by dipping the person's fingers in ink, one at a time, and recording these on the National Fingerprint Form. The fingerprint expert then compared the **ridge patterns** from the fingerprints on the form with those taken from the scene.

Scanning a fingerprint

Nowadays police use LiveScan – a scanning device – to scan a suspect's fingerprints. From a **LiveScan** terminal anywhere in England, Scotland or Wales, the ten fingerprint impressions can be transmitted electronically to the National Fingerprint Database.

If fingerprints from a crime scene are scanned, within seconds computer software called IDENT1 provides a list of possible suspects from those on the database. It's then left to the fingerprint expert to make the decision as to which of the suspects left the fingerprints at the crime scene.

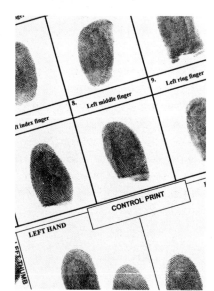

The National Fingerprint Form

An example of a fingerprint pattern

...national fingerprint form

M3

Think about

Although it is known that they were not involved in the attack, the fingerprints of the bar staff at the pub were also taken. Why?

Think about

UK fingerprint experts went to Thailand following the tsunami in December 2004. Fingerprints were taken from bodies of disaster victims. Explain how the team would have used databases and items belonging to missing persons from the UK to identify the victims.

Discuss

*Some schools use fingerprint patterns for identification purposes for entry systems into the school or library. In the future, fingerprint and other **biometric information** may be used on **national identity cards**. Discuss the advantages and disadvantages of fingerprinting the entire population of the UK.*

Fingerprint patterns

When comparing fingerprints from the crime scene and those from a suspect, first of all the fingerprint expert looks at the overall patterns. Fingerprint patterns fall into three types called **loops**, **whorls** and **arches**.

 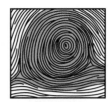 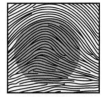

| Loops are the commonest type | Whorls | Arches are the least common type |

From fingerprint patterns alone, it's not possible to make a positive identification. But it may be possible to narrow down the list of suspects or eliminate some people.

D3

Fingerprint ridges

If the patterns are found to be in agreement, the fingerprint expert can move to the next stage. He looks at characteristics of the ridges – are these of the same type, and are they in identical positions?

There are seven types of ridge characteristic.

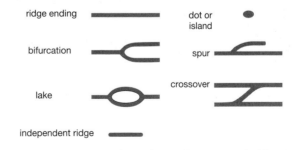

Fingerprint ridge matches have been the most reliable evidence for proving, beyond reasonable doubt, that someone was present at a crime scene.

Testing for body fluids

Is it blood?

Sam hasn't been seen for several days. A relative enters her flat and sees spots of what appears to be blood on the kitchen floor. But is it blood? A forensic scientist can test the spots before samples are collected and sent to the lab for analysis.

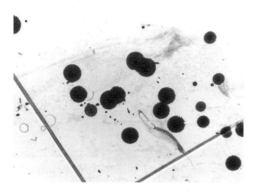

P4 Body fluids at a crime scene

Blood is a type of **body fluid**. Body fluids, and stains made from body fluids, are often found at the scene of a violent crime. It's the job of the forensic scientist to determine what the body fluid is, and from whom it came.

Three types of body fluid are important to the forensic scientist.

- **Blood** contains red and white blood cells, along with cell fragments called platelets, suspended in a light-coloured liquid called plasma.

- **Semen** may be found at a crime scene or on a victim if sexual contact has taken place. Semen contains spermatozoa, or sperm, in a thick white fluid. The fluid is composed of food substances for the sperm, mucus, enzymes and hormones.

- **Saliva** is sometimes found on clothing at a crime scene, or on objects such envelopes or labels. Saliva is made up mostly of water, but it also contains the enzyme amylase, sodium hydrogencarbonate and mucus.

When you lick an envelope you leave traces of saliva

Know more

Body fluids from a crime scene pose a serious risk during their collection, their analysis and their use in court. The main hazards are the possible presence of the viruses hepatitis B and HIV.

...forensic testing blood

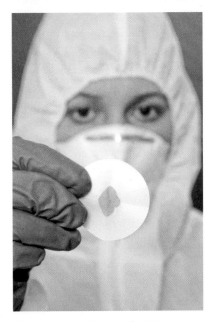

A forensic scientist uses the KM test to look for blood

Testing for blood at the crime scene

Any stain at the crime scene that may be blood should be tested. A forensic scientist called to the scene will test a scraping of one of the spots using the **Kastle-Meyer (KM) test**. A deep pink colour will suggest that blood is present. But even with this positive result, it is not absolutely certain that the spot *is* blood. The KM test is therefore called a **presumptive test**. With a positive result, and the other visual evidence available, the forensic scientist can *presume* that blood is present, but can't be sure at this stage.

What's more, the KM test doesn't tell the forensic scientist whether the substance is likely to be human blood, or blood from another animal. But the test can confirm that a substance *isn't* blood.

Enhancing bloodstains

Another presumptive test for blood uses a chemical called **luminol**. It is very useful for enhancing any bloodstains that a criminal may have tried to remove by wiping a surface or washing clothes. It can detect blood that has been diluted ten million times.

Testing for other body fluids

Any item suspected as having semen on it is tested in the lab. The forensic scientist uses the **acid phosphatase test**. This detects an enzyme that's present in semen. The sample turns purple if semen is present.

To test a sample for saliva, the *Phadebas*® test is used. This relies on the fact that saliva contains the enzyme amylase, which breaks down starch. *Phadebas*® is a starch polymer dyed blue. In the presence of amylase, the blue dye is released.

Phadebas® *tests for saliva*

Know more

Presumptive tests can give false positive results. The KM test can suggest that blood is present when it comes into contact with some vegetable materials, such as potatoes.

Testing blood samples

Your assessment criteria:

P4 Carry out experiments to analyse biological, chemical and physical evidence from the crime scene

> Practical

M3 Describe the patterns found from the evidence and make connections

D3 Explain the patterns found from the evidence and make connections.

Whose blood is it?

Sam hasn't been seen for a few days. A forensic scientist carries out a quick presumptive test on stains on the floor of Sam's flat. She gets a positive result for blood. With the evidence available, she decides to send a sample to the lab.

P4

A conclusive test for human blood

The forensic scientist at the lab needs to prove that the sample is blood, and that it's of human origin rather than from another animal. He uses a **conclusive test**.

The **precipitin test** is a conclusive test used to identify human blood. A drop of the sample is mixed with an **anti-human serum** (plural sera). After a minute, a precipitate (solid deposit) is produced if the sample is definitely human blood.

Blood group tests

The next step could be to find out who the blood came from. To start this, the blood group of the sample could be determined. The four blood groups are A, B, AB and O. The forensic scientist tests the blood with two different anti-sera.

Blood group	Percentage of UK population
A	42
B	9
AB	3
O	46

Blood groups

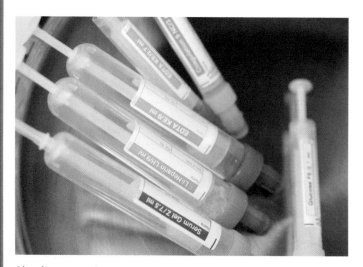

Blood group testing

🔍 ...forensic preciptin test

P4

Blood groups can help to identify the person that the blood came from. But many people have the same blood group. In the UK, blood groups O and A are the most common.

The other important blood group system is the **rhesus blood group system**. People fall into one of two groups. They're either rhesus-positive or rhesus-negative. Most people in the UK (around 84%) are rhesus-positive.

If the missing victim, Sam, is known to have had a rare blood group, such as AB rhesus-negative, and blood of this type is found at the scene, this gives a very strong suggestion that the blood spots come from her.

For common blood groups, blood group evidence may not be of much use in identifying a person. But when blood groups *don't* match, people can be eliminated from the investigation.

Know more

We have many enzymes in our blood. Enzymes are proteins that regulate the rate of chemical reactions in the body. Some show certain variations in the human population. The detection of particular enzymes in a sample of blood, and the type of variation present, can help to identify who the blood came from.

Immunological tests

M3

The conclusive test for blood, the precipitin test, is an example of an **immunological test**. The anti-serum contains **antibodies** against proteins in human blood. When (and only when) it meets with human blood proteins, it causes the proteins to clump together.

How the blood group tests work

D3

Our blood groups are the result of chemicals, called **antigens**, on the surface of our red blood cells. These are determined genetically.

The forensic scientist tests the blood with two anti-sera – anti-A serum and anti-B serum. These react with A or B antigens. When a matching antigen and anti-serum meet, the blood cells clump together. They are said to be **agglutinated**.

Research

Find out how human anti-serum is made. Many forensic laboratories carry anti-sera for a range of common animals. How are these made?

Blood group	Antigens on the surface of the blood cells	Reaction of blood with anti-A serum	Reaction of blood with anti-B serum
A	A	agglutination	none
B	B	none	agglutination
AB	both A and B	agglutination	agglutination
O	none	none	none

...laboratory human anti-serum

Analysing blood patterns

How did the blood get there?

A burglar breaks into a remote farmhouse. The owner hears the intruder and goes downstairs. He attacks the burglar, hitting him on the side of the head with a walking stick. The burglar escapes, bleeding.

Your assessment criteria:

P4 Carry out experiments to analyse biological, chemical and physical evidence from the crime scene

Practical

M3 Describe the patterns found from the evidence and make connections

D3 Explain the patterns found from the evidence and make connections.

Know more

A droplet of blood travels through the air as a perfect sphere. It only breaks up when it comes into contact with an object.

P4 Blood at the crime scene

Forensic scientists examine and record the **blood patterns** at crime scenes. Blood could be from the victim and/or the attacker, it could be on the floor and walls, on clothes or on weapons. **Transfer stains** may also have been left, produced when blood was transferred to clean surfaces by people's shoes, hands or hair, or a cloth that was used to wipe a weapon.

Bloodstains can be made by falling blood, projected blood or smeared blood. The forensic scientist examines and carefully measures the shape and size of the bloodstains. These can give information on:

- the volume of blood lost from the body
- the force of any impact that led to the loss of blood
- the direction in which the blood was travelling
- the angle at which the blood hit a wall or floor.

This evidence will reveal much about how the crime took place. It could also confirm or dispute eyewitnesses' or victims' versions of what happened. With careful measurements, it may be possible to reconstruct the crime.

Know more

The surface that the blood hits is of major importance when interpreting bloodstain evidence. The shape of the bloodstain will be very different on hard and soft surfaces. Carefully controlled experiments may have to be carried out before valid conclusions can be drawn.

...blood spatter diagrams

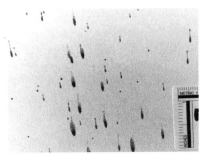

Blood spatter from a medium-force impact (for example with a weapon such as a cricket bat). The spots might be 1 to 8 mm in diameter

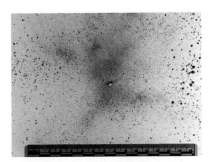

Blood spatter from gunshot. The spots of blood are usually less than 1 mm across

The volume of blood

The amount of blood left at a crime scene can tell us a lot about the incident. An adult male has 5 to 6 dm^3 of blood in their body, and an adult female 4 to 5 dm^3. When a major injury occurs, large volumes of blood can be lost. When an artery in the body is severed, the blood is released under high pressure.

The force applied

When Sam had a nosebleed in her flat, the spots hit the floor with a low force, under the influence of gravity.
The spots on the floor were more or less circular. From these spots, we can also get a good indication of the height from which the blood fell.

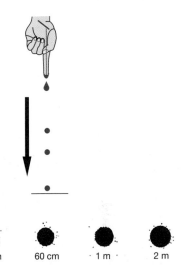

| blood falling from height of | 15 cm | 30 cm | 60 cm | 1 m | 2 m |

The diameter of blood spots (for the same volume of blood) increases as the height from which they fell increases

On the other hand, if the body is hit with force, blood is projected with speed. The bloodstains produced usually form a pattern called a **blood spatter**. Blood gets scattered in drops over a wide area. The shape and size of blood spatters is very important to the forensic scientist.

In the case of a high impact, such as when a gun is fired, the blood travels at a much greater speed and a fine mist of blood droplets is produced.

Think about

Using evidence from several spots, think about how the position of the source of the blood could be pinpointed. Draw a diagram.

The shape of the stain shows where the blood came from

The direction the blood was moving in

It is possible to get information about the direction the blood was travelling in by looking at the shape of the stain.
The pointed end of the stain always faces the direction that the blood was travelling in.

From the dimensions of a bloodstain, the forensic scientist can even work out, using simple maths, the angle at which the blood hit the floor.

 ...crime scene blood direction

DNA profiling

Your assessment criteria:

P4 Carry out experiments to analyse biological, chemical and physical evidence from the crime scene

> **Practical**

M3 Describe the patterns found from the evidence and make connections

D3 Explain the patterns found from the evidence and make connections

Body fluids left at the crime scene

A burglar attempts a robbery at a farmhouse and is injured as the owner attacks him. Bloodstains have been left. To be certain in the identification of the criminal, the forensic scientist needs to look at DNA in the blood.

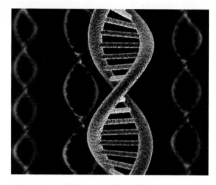

The DNA molecule; is found in the nucleus of almost every cell in our bodies

P4

DNA evidence

Body fluids such as blood, saliva and semen all contain cells with DNA in their nuclei. Body fluids left at the crime scene will therefore be a good source of DNA. DNA can also be extracted from hair with part of the follicle attached, from skin and dandruff, from bones and teeth, and possibly from other items such as cigarette butts and tissues at the crime scene.

DNA is inherited, so there are family similarities, but each person's DNA is *unique* (except for that of identical twins). It can provide crucial evidence in a criminal investigation.

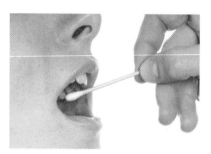

A swab is used to take cheek cells from a suspect's mouth. The cheek cells are a good source of DNA

Producing a DNA fingerprint

To analyse the DNA, the forensic lab needs to produce a DNA fingerprint. DNA is extracted from the blood sample and then the DNA molecules are cut into fragments. These DNA fragments carry an electrical charge. This means that they will move in an electric field. The fragments are put into small cavities at one end of a gel, and a voltage is applied across the gel. This is called **electrophoresis**.

The DNA fragments move through the gel. The resulting pattern of bands on the gel is a DNA 'fingerprint' or **DNA profile**.

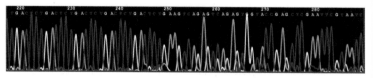

Part of a modern DNA fingerprint. The electrophoresis gel is scanned to produce a series of peaks

> ### Know more
>
> *If body fluids from a crime scene are to be used as DNA evidence, the equipment used for their collection, such as swabs or scalpels, or water for dried stains, must be sterile to avoid contamination.*

🔍 ...DNA fingerprinting

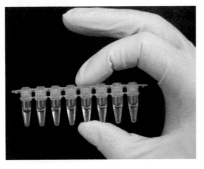

A strip of eight PCR tubes

Copying DNA

P4

Often, only very small amounts of DNA are recovered from crime scenes. But the forensic scientist must have enough DNA to work with. An enzyme called **DNA polymerase** is used in a process to copy the DNA sample. In around 2½ hours, around a billion exact copies of the DNA to be analysed can be produced. The process is called the **polymerase chain reaction** or **PCR**.

Research

The National DNA Database (NDNAD) contains DNA fingerprints of over 4 million people. Find out whose DNA is added to the database and how long it remains there.

Discuss

A forensic scientist is able to obtain a match for nine standard points, or loci, of the suspect's DNA fingerprint. He estimates that the probability of the two samples coming from different people is 1 in 1 billion, or 1 in 1 000 000 000. The population of the UK is around 60 million. Discuss what this evidence tells us.

Analysing DNA fingerprints

M3

The forensic scientist compares the DNA profiles of suspects with DNA from the crime scene.

Research

Police forces in the UK keep crime scene evidence for many years. DNA evidence can be used to solve 'cold cases', where the case remains unsolved after a number of years. Research a cold case that DNA fingerprinting has been able to solve.

What does a matching DNA fingerprint mean?

D3

If there's a match between two DNA fingerprints at eleven standard points, or *loci*, then the probability of the two DNA samples having come from *different* people would be 1 in a trillion. (That's 1 in 1 000 000 000 000.)

But it's not always possible to look at all eleven points. If the DNA sample is incomplete or badly preserved, the forensic scientist will have fewer points to investigate.

Forensic entomology

Your assessment criteria:

P2 Demonstrate the most appropriate methods to record and preserve evidence from the crime scene

M3 Describe the patterns found from the evidence and make connections

D3 Explain the patterns found from the evidence and make connections

Insects are found on a body

A man walking in a remote part of a forest finds the body of a young woman lying in the undergrowth. There are flies buzzing around the body, and many maggots feeding on it. When a SOCO arrives at the scene, he calls straightaway for an expert on insects and forensic science – a **forensic entomologist**.

P2 Collecting and recording the insects

When the forensic entomologist arrives, he (or she) records the number and types of insects on and around the body, and the stage they are at in their development or **life cycle**.

He then collects insects from the body, and from the surrounding earth up to a few metres away from it. He makes a careful note of the habitat and the weather conditions.

He kills some insects of each type using a chemical agent, and preserves them in alcohol.

Forensic entomology sample

The rest of the insects he has collected are taken back to his lab alive. The **specimens** may provide important information to the investigation.

- There is a set sequence of types of insect that lay eggs on or feed on a dead body, so a good estimate can be made of when the person died.

- Careful identification of the insects and knowledge of their habitats gives information about whether or not the person died where the body was found.

> **Know more**
>
> *Insects go through four distinct stages in their life cycle.*

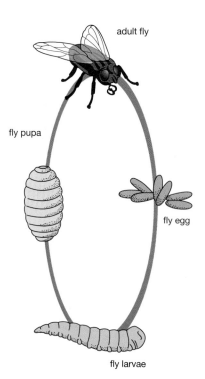

The life cycle of a fly

adult fly

fly pupa

fly egg

fly larvae

...insect life cycles

Hide beetles come to feed as the body breaks down

The sequence of insects on a body

M3

If a body is left unburied, **blowflies** (such as bluebottles) arrive in minutes and lay eggs on it. They are attracted by body fluids and by the gases released by the early stages of the body's decomposition (hydrogen sulphide and ammonia). They lay their eggs in the mouth, nostrils, eyes, and in any wounds.

After a few days, the haemoglobin in the body breaks down and more gases are produced. The body becomes bloated. **Flesh flies** arrive.

The flies contribute most to the breakdown of the body. But a **range of other insects** will also be involved, depending on conditions and location.

After about a month, the body collapses. As fats break down, **various beetles** arrive, along with **insects that are predators on these beetles**.

After three to sixth months, the breakdown of the body's protein attracts **cheese flies** to feed.

As the body becomes dehydrated, **certain moths and mites** arrive to feed.

Think about

The coffin fly is associated with bodies that have been buried for at least a year. It can give indications of how long the body has been buried. But what will it tell us if larvae of the coffin fly are found on a body left exposed or lightly covered with leaves?

When and where did the death occur?

D3

The forensic scientist measures the size of the different types of maggots collected. This can be a very good indicator of their age. He also rears the maggots he has collected under carefully **controlled conditions** (for example, temperature and humidity). He knows the length of the life cycles of different flies under these conditions. So when the adult flies emerge, he can work backwards to calculate when the eggs were laid.

In many crimes, a body is moved from the place of murder and left somewhere else. A body found in a rural area may have insect species feeding on it that are only found in urban areas – a clear indication that a body has been moved.

🔍 ...coffin fly and forensics

Identifying remains

Human skeleton discovered

The new owners of a house are having it renovated. Workmen discover a skeleton beneath a slab of concrete in the garden. A forensic anthropologist is brought in to investigate.

P4

Forensic anthropology

Forensic anthropology is a specialist area of forensic science, involving human skeletal remains. Examination of a skeleton can give the investigating team information about the height of the person, their sex and racial origin. It may also reveal injuries to the body, either before or after death.

A skeleton can also provide some information about the age of the person when they died, depending on whether or not the bones have finished growing. It may also be possible to tell the age of the skeleton itself.

Forensic odontology

Teeth are the most durable part of the human body. They can even survive extremes of temperature, for example when a body is burnt. The structure, number and pattern of teeth in the jaw, and special features such as cavities and crowns, are examined by a **forensic odontologist**. Teeth can identify a specific person.

Tooth structure can provide information about the age of a person. The enamel becomes worn as the person ages. The root also becomes more flattened.

Dentists keep detailed records of the layout and structure of their patients' teeth. A person's unique **dental records** will show if a person has teeth missing, and where these gaps are on the jawbone. It will also show up any cavities in the teeth, loose teeth, fillings and crowns.

As well as helping to identify human remains, dental records can help in investigations where bitemarks have been left, for instance in a person or piece of food. However, if a person has a perfect set of teeth, a dental record may not be of much use.

Know more

The time for a dead body to change to a skeleton varies. In temperate climates (such as our own), the decomposition can take about two years. In tropical climates, decomposition may be complete in only a few weeks. If a skeleton crumbles easily, then it's probably over 100 years old.

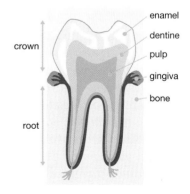

The structure of a molar tooth

 ...age of skeleton

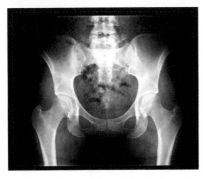

An x-ray showing the pelvis of a female human

Information from the skeleton

M3

The shape and size of the pelvis, skull and long bones indicate whether a skeleton is male or female, and its racial origin.

The skeleton can also provide important information about disease, such as **osteoporosis** (brittle bone disease). If the forensic anthropologist finds any features of the skeleton that might be significant, investigators can check medical records. Old bone injuries such as fractures may be important in the identification of the remains. Non-human structures, or artefacts, can also be found with the skeleton. These might be surgical implants that the person was given, such as pins to hold joints together. Or they could include pieces of jewellery, such as studs, rings and necklaces.

The forensic anthropologist may also pick up signs of violent injury, such as holes in the skull or stab wounds.

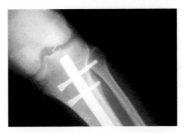

An X-ray of a joint from a person's medical records showing a pin in the joint

Forensic anthropologists can often match stab wounds to the type of knife used

When a skeleton is complete, it is easy to estimate the height of the person. But when it isn't, the forensic anthropologist can calculate the height from the length of the skeleton's thigh bone (femur) using a mathematical formula.

Facial reconstruction

Facial reconstruction gives an idea of the appearance of the person when he or she was alive, so may help with the investigation. Using the skull's characteristics, layers of modelling clay are built up on a cast of the skull.

Think about

When there's a mass disaster, such as the destruction of the World Trade Center in 2001, forensic odontology can be used to identify people when no other means of identification is possible. What techniques could investigators use to identify victims? What difficulties would they encounter?

A face being reconstructed

DNA in bones and teeth

D3

If a person has died recently, DNA evidence can often be obtained from the long bones of the body. The roots of teeth – the dentine and pulp – are also an excellent source of DNA, even if they are very old. DNA has been extracted from teeth that are hundreds of years old in historical and archaeological investigations.

...identifying victims by teeth

Marks and impressions

Forced entry

A house has been burgled. When the SOCO arrives at the scene, there is evidence to suggest that a tool has been used to force open a window. There are shoeprints in the soft soil around the window.

Shoeprint

P2 Making casts

The shoeprints left at the scene could be a major piece of evidence. And so could the **impression** in the UPVC window frame, left by some kind of tool used to lever open the window. The SOCO first photographs the shoeprints and tool marks.

The SOCO then makes **casts** of both types of impression, as permanent records. He (or she) places a casting frame around the shoeprint. He mixes some of the casting material and pours this into the shoeprint. It is prevented from spreading further by the casting frame. The cast is made with a type of fine plaster called dental stone, which is also used by dentists. This sets hard and it shows up fine details in the shoeprint.

After about 30 minutes, when the cast is set, it is removed. The cast is allowed to dry out for a further day or two, then it is cleaned carefully with a brush.

For recording the toolmark impressions in the window frame, the SOCO uses a type of silicone rubber as the casting material. When it sets, the cast is flexible, so can be removed from a UPVC or window frame without it breaking.

Evidence from a suspect

If a tip-off leads investigators to a suspect, the person's shoes and tool set are taken away for examination and tests.

Investigators make **test prints** of the shoes. They do this by covering the sole with water-based ink and then pressing onto a sheet of acetate. This can then be compared with the cast taken from the crime scene.

Know more

Red 'snowprint powder' is applied to shoeprints in snow to make the details visible in photographs.

Know more

If the soil is soft, it may be necessary to seal the shoeprint first. Otherwise, it might simply break up when the casting material is poured in.
The SOCO sprays the shoeprint with hairspray. A waxy powder is used on snow.

🔍 ...crime scene and shoeprint

M3

Research

Find an online shoeprint database. How many types of treads does it include? Is the database easy to use?

What information do casts provide?

The shoeprint cast will provide the investigators with information about shoe size and type of shoe worn. Trainers, for example, often have characteristic tread patterns that make identification of the make and style of shoe relatively easy. Shoeprint databases are available to speed up the process of identification.

The cast of the toolmark will also identify the type of tool that made the mark, for instance a crowbar or screwdriver.

spatula

silicone is applied to the impression

Making a cast of a toolmark

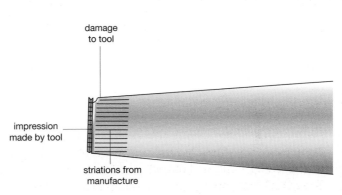

damage to tool

impression made by tool

striations from manufacture

Matching striations on a screwdriver to marks in an impression

D3

Shoeprint and tool impressions can be unique

Sometimes shoe treads can be *unique* to one shoe. They are likely to show signs of wear. The way an individual walks often wears down one side of the shoe more than the other. There may also be damage on a sole, or imperfections from the manufacture of the shoe.

Tools are also often unique. Marks, or striations, are made on them as part of the manufacturing process. And during their normal use, they can be become worn or damaged. Close comparison of casts of tools and casts of impressions can sometimes prove, beyond reasonable doubt, that a particular tool was used in the crime.

Think about

The team investigating the forced entry have made casts of the screwdrivers from a suspect's shed. Explain why they did this. Why didn't they try to fit each screwdriver into the impression on the window frame until they found one that was a fit?

🔍 ...shoeprint database

Identifying fibres

Crime and fibres

A person has been mugged in the street. The SOCO finds some fibres on his jumper. They are a different colour from his jumper and appear to be of a different texture.

The forensic scientist will analyse the fibres.

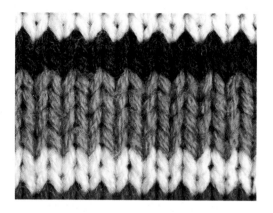

P4

Collecting fibres from the crime scene

The SOCO cuts a piece of lifting tape and applies this to the victim's jacket. The tape is lifted and then stuck to a piece of acetate film, which is labelled and stored.

Tape-lifting is used to collect fibres in the majority of cases. But picking the fibres off a garment by hand is also an alternative. In some cases, a special vacuum device is used. Air is sucked through a paper or fabric filter, and the fibres collected can be recovered from the filter.

Examining fibres

The forensic scientist can see certain features of the fibres, such as the colour, just by examining them with the naked eye. But he (or she) will need to use a microscope to examine them fully.

He first removes a fibre from one of the acetate films. Sometimes a solvent is needed to dissolve the adhesive. He then mounts one of the fibres on a microscope slide.

He examines the slide with a standard **light microscope**. Adjusting the direction of the light helps him to see the surface of the fibre in detail.

Angora wool fibres seen with a light microscope

Know more

*The technique of **scanning electron microscopy (SEM)** is used to give detailed images of the surface structure of fibres. The fibre sample is coated with a heavy metal, then the sample is bombarded with electrons which are collected and displayed on a screen. This gives an almost 3-D image of the material.*

...micrograph of silk fibre

In a wool fibre the scales of the hair cuticle can be seen

A silk fibre is made up of two filaments

A cotton fibre is ribbon-like, with twists at regular intervals

The fine structure of fibres

P4

Fibres fall into two groups – natural and synthetic. **Natural fibres** are from animal and plant sources. They include hair from sheep (wool), goats (mohair and cashmere) and rabbits (angora). Silk is produced by unravelling the fibre from the cocoons of the silk moth.

Fibres of animal origin are made from protein. Plant fibres include cotton and linen, both made from the chemical cellulose.

All of these natural fibres, when viewed microscopically, have very distinctive features and so are easily identifiable.

Synthetic fibres are much more uniform, and difficult to identify with the microscope.

Chemical techniques for analysing fibres

As well as looking at fibres microscopically, chemical tests can be carried out. A small sample of fibres is placed in a drop of chemical solution (for instance, sodium hypochlorite or sodium hydroxide). The behaviour is noted – fibres may dissolve.

Chemical analysis can also reveal the dyes used to give colour to the fibre. These are often easily removed using a solvent, and can then be compared using **chromatography**.

Think about

Why must the forensic scientist make sure that control fibres taken from a suspect's garment are representative of the structure of the garment?

Know more

Infrared spectroscopy *is a technique for analysing the chemical nature of fibres. This is very useful in the identification of synthetic fibres, which are very similar when viewed with a microscope. FTIR (Fourier transform infrared spectroscopy) analyses a fibre sample in around a second and can distinguish the different types of synthetic fibre.*

Comparison with control fibres

M3

The validity of fibre evidence relies on the careful comparison of fibres from the crime scene or victim with those from the suspect's garment. When a suspect is apprehended, control fibres must be removed from the suspect's garment for examination and tests. The use of a comparison microscope, as with analysing hair, can help in the matching procedure.

How reliable is fibre evidence?

D3

Great care must be used when interpreting evidence from fibres. Fibres are transferred very easily and may be passed on several times. They may also stick to a garment for some time.

The combination of the type of a fibre sample and the dyes in it may be quite rare. If there is then a match between fibres from the crime scene and those from a suspect's garment, and there is other supporting evidence, it may be reasonable to suggest that they came from the same source.

...high-performance liquid chromatography

Analysing glass

Ram raid at the bank

A van is driven at high speed into a bank, smashing the front glass window. A group of masked men jump out of the van and wrench the cash machine from the wall. The men drive off with the cash machine and the van is later found abandoned in the countryside.

P4

Collecting evidence from the crime scene

A team of SOCOs collects pieces of glass from the bank. A separate SOCO team collects tiny fragments of glass from the abandoned van, and removes the van's headlamp, which has been broken in the ram raid.

Analysing glass

When glass is smashed, it breaks into many pieces of different sizes. There may be some large pieces, but it is likely that a wide area is showered with tiny **fragments**.

Forensic scientists are sometimes able to fit pieces of glass together, for example they might find a piece of the van's headlamp glass on the bank and check that it fits the broken lamp. This evidence would be very strong.

The refractive index of glass

One of the most useful tests carried out on glass is to find out its **refractive index**. When rays of light pass from air through a piece of glass, they are bent, or refracted, by the glass. The amount they bend depends on the value of the refractive index. (See Unit 5 for more on refraction.)

Since fragments of glass from a crime scene are often very small, or irregular in shape, forensic scientists need to use a specialist technique to measure the refractive index. One technique involves putting the fragment of glass in some silicone oil.

The oil is warmed up slowly. The refractive index of the silicone oil falls as the temperature is raised. But the refractive index of the glass changes very little.

Refraction causes this straw in a glass of water to appear broken. Light rays bend towards the normal (an imaginary line at right angles to the surface of the glass) when travelling from air to glass

 ...forensic glass analysis

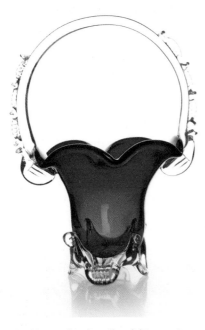

The blown glass handle of this vase is less dense than window pane glass

P4

The glass sample is viewed with a microscope. As the oil gets warmer, there comes a point when the refractive index of the glass and the silicone oil are the same, and the edges of the glass fragment can no longer be seen.

GRIM3 analysis

Measurement of the refractive index is now automated. Using a device called GRIM3 (Glass Refractive Index Measurement), an image of the sample can be seen on a computer screen. At the matching temperature, the refractive index of the glass is calculated using stored data.

Other properties of glass

The forensic scientist can also measure other physical and chemical properties of glass that may help in its identification. The combination of chemical **elements** in glass depends on the type of glass and how it was made. Detailed analyses of the composition of glass can be carried out using instrumental techniques. Different types of glass also have different **densities**.

Know more

To measure the density of a fragment of glass, the forensic scientist places it in a liquid. He changes the density of the liquid by adding small quantities of a second liquid, until the fragment doesn't sink or float. The densities of the liquid and the fragment are then identical.

M3

What the refractive index can tell us

Different types of glass each have a slightly different refractive index (see the table). So it's possible to match samples of glass collected from a crime scene or suspect.

Glass	Refractive index
headlamp	1.47 – 1.49
television	1.49 – 1.51
window	1.51 – 1.52
bottle	1.51 – 1.52
lead crystal	1.60 – 1.70

Think about

The SOCO took a piece of glass that was still in the window frame at the bank. This was used by the forensic scientist as the control sample. Explain what is meant by a control sample.

D3

Making comparisons

The SOCO needs to look for glass fragments on the suspect's clothes. He does this by shaking them individually over a sheet of paper. He also searches for tiny glass fragments in the person's hair by combing it, again over a sheet of paper. These fragments are then sent to be forensically compared with fragments from the scene of the crime.

...refractive index of windscreen

Analysing paint

The hit and run

A van that is out of control mounts the pavement and knocks down a pedestrian. The van also hits a wooden gate before speeding off. An eyewitness notes the colour of the van and notes down part of the registration number. The SOCO finds a flake of paint next to the gate and sends it to the forensic scientist for analysis.

P4

Examining paint

The colour of a paint sample is an obvious way of matching it to a source. Over the years, car manufacturers have used thousands of different colours and shades of car paint. Luckily they keep records of the colours used during different manufacturing periods.

A forensic scientist compares a paint sample from the crime scene with known types of car paint

The forensic scientist can also examine the structure of a sample of paint with a microscope. Paint on cars is built up as a series of layers, and newly manufactured cars have a specific **sequence**.

Scanning electron microscopy (SEM) can be used to reveal greater detail of both the paint's surface structure and the layers that make up the paint sample.

Know more

Paint flakes are cut using the knife of an instrument called a **microtome** *to obtain sections with clean edges so that the layers can be observed easily.*

Car paint seen with scanning electron microscopy

 ...crime scene paint samples

Chemical analysis of a paint sample

P4

Paints are usually made of **pigments** suspended in a **binder**. The pigments give the paint its colour. Paint pigments can be either inorganic or organic chemicals. The binder holds the particles of pigment together in a coat of paint. It is usually a polymer.

The complex formulation of paints means that their chemical analysis is also complicated. The forensic scientist uses different techniques to analyse inorganic and organic components of the paint sample.

A special type of **gas chromatography**, called pyrolysis gas chromatography, is used. This gives a very detailed analysis, called a **pyrogram**, of the polymers in a sample of paint as small as a few micrograms.

Think about

A suspect is found and all the evidence fits, except for his car colour. The forensic scientist examines a paint sample from the scene of the hit-and-run, and a sample from the suspect's car. The results are shown below. What do the results suggest?

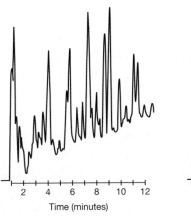
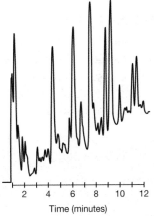

Time (minutes)   Time (minutes)

Pyrogram of two paint samples from different cars

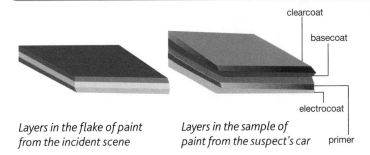

Layers in the flake of paint from the incident scene

Layers in the sample of paint from the suspect's car

clearcoat
basecoat
electrocoat
primer

How the paint samples help the investigation

M3

After an incident such as a hit-and-run, if some of the car registration number is known, investigators look through the **DVLA car registration databases** looking for possible suspects. They visit these in turn. Eventually, a suspect may be found who can't account for his movements on the day. The car's paintwork is then examined.

🔍 ...vehicle paint layer analysis

Testing for drugs

A drug overdose

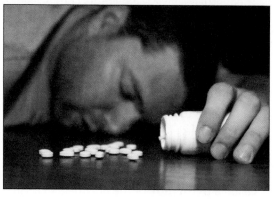

A man is discovered dead in his flat. As the SOCO makes her search of the flat, she finds a powder in a small plastic envelope. It could be a recreational drug. She packages up the evidence and sends it to the forensic scientist. The body is taken to the mortuary where the forensic toxicologist will carry out tests on it.

Your assessment criteria:

P4 Carry out experiments to analyse biological, chemical and physical evidence from the crime scene

> Practical

M3 Describe the patterns found from the evidence and make connections

D3 Explain the patterns found from the evidence and make connections

P4 — Testing for drugs

The **forensic toxicologist** first uses a quick test on the suspected drug. He uses **Marquis reagent**. The reagent goes through a sequence of colour changes that can give an indication of the drug that's present.

Further tests are needed though to confirm the identity of the substance. **Thin-layer chromatography (TLC)** is similar to the paper chromatography you've carried out in school. But instead of using chromatography paper, the sample is put onto a TLC plate. This is a sheet of glass coated with a thin layer of powder, such as silica gel.

The forensic scientist dissolves some of the suspected drug in a suitable solvent, then puts a spot of the solution on the TLC plate. When the spot is dry, he places the TLC plate in a chromatography tank containing a mixture of solvents. After one hour, he removes the TLC plate and quickly marks the height reached by the solvent. When it's dry, he examines the TLC plate under ultraviolet light.

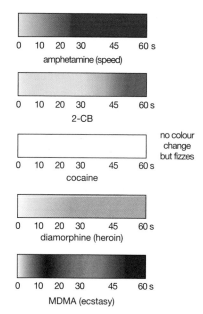

Colour changes obtained for some drugs using Marquis reagent

Know more

TLC can provide very good evidence but not conclusive proof. For this the forensic scientist must use either gas chromatography or high-performance liquid chromatography, linked to an instrument called a **mass spectrometer**. *The technique is called GC-MS or HPLC-MS. The mass spectrometer positively identifies the chemicals in the sample.*

🔍 ...Marquis reagent colours

Think about

A forensic scientist has identified a substance, by TLC, as cocaine. But other spots appeared on the chromatogram. Sometimes other chemicals are added to cocaine to increase the amount of the drug for sale. How can the TLC results help the forensic scientist to identify these?

Think about

Suggest what the unknown substance is in the chromatogram shown here.

Research

Find out more about the use of gas chromatography (GC-MS) and high-performance liquid chromatography (HPLC-MS) in forensic applications.

Think about

The forensic toxicologist found a blood concentration of cocaine of 2.50 mg/dm³. He refers to the table of reference figures given here. Explain what his results suggest.

Analysing results of TLC

As well as the unknown substance, the forensic scientist also places pure samples of different drugs on the TLC plate. After one hour, the different drugs have risen to different heights on the finished TLC plate, or **chromatogram**. By comparing the heights reached by the unknown substance and the samples of known drugs, the forensic scientist can suggest the identity of the suspected drug.

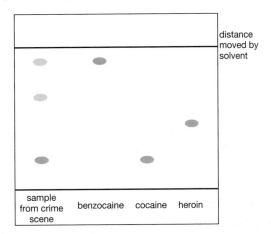

How does TLC work?

When the TLC plate is placed in the chromatography tank, the solvent rises up the coating on the glass. As the solvent moves past a drug on the plate, it dissolves it. Some of the drugs are more soluble in the solvent than others. And some drugs are more attracted to the silica gel than others. The drugs therefore move to different points on the finished TLC plate, forming a chromatogram.

Did the drug cause the person's death?

All the evidence may point to the person having used cocaine. But the investigators also need to know whether use of the drug actually caused the death.

Using GC-MS, the forensic toxicologist finds the concentration of cocaine in the body's blood. He compares this with reference figures.

range following normal dose, mg/dm³ blood	range following toxic dose, mg/dm³ blood	range following fatal dose, mg/dm³ blood
0.05 – 0.30	0.25 – 5.00	1.00 – 20.00

🔍 ...drugs in bloodstream

Identifying poisons

Was he poisoned?

A man was discovered dead in his flat. When the SOCO carried out her search, she found an unlabelled plastic packet close to the body, containing some white crystalline powder.
She sends it to the forensic scientist for identification.

P4

Poisons

Powder found near a body could be just a household chemical. Or it could be a poison. Many different chemical compounds are used as poisons.

- **Corrosive poisons** include hydrochloric, nitric and sulphuric acid, and sodium hydroxide.
- Certain metals and metal **ions** are poisonous, including lead, mercury and thallium.
- Many non-metal ions, such as cyanide, are poisonous.
- **Organic compounds** such as pesticides, and toxins from plants and animals, are also used as poisons.

Testing for metal ions

You can carry out simple tests to detect metal ions in the school lab. Some metal ions, or **cations**, will glow a distinctive colour if put in a Bunsen burner flame. **Flame testing** is a qualitative test – it tells you whether a chemical is in a sample, but not how much of it.

Carrying out a flame test

Some chemical tests will also show up the presence of metal ions. The addition of sodium hydroxide, for instance, to a test solution, will sometimes produce a precipitate (solid matter). This is a **precipitation reaction**. The colour of the solid can help in the identification of the metal ion.

Precipitation reactions can also be used to detect non-metal ions, or **anions**.

Know more

Ions are electrically charged atoms. A cation is an atom that has lost one or more electrons. It is positively charged. An anion is an atom, or group of atoms, that has gained one or more extra electrons. It is negatively charged.

 ...chemical poisons

P4

Adapting the flame test for forensic work

Forensic scientists use instrumental methods that work in a similar way to flame tests. **Atomic emission spectroscopy (AES)**, or an advanced technique called **AES-ICP** that uses a very hot plasma flame, can detect metal ions in a sample in very low concentrations. Also, unlike flame tests, it doesn't matter if other ions are present – a number of metal ions can be analysed at the same time. This technique also measures the amount of a metal ion in a sample – it's a quantitative test. This is very important to the forensic scientist when looking for a poison.

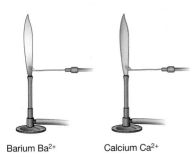

Barium Ba^{2+} Calcium Ca^{2+}

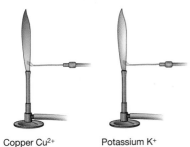

Copper Cu^{2+} Potassium K^+

Flame test colours

M3

The results of simple qualitative tests

Precipitation tests

Cation tested for	Solution added to test sample	Colour of precipitate	Name of precipitate
copper(II) Cu^{2+}	sodium hydroxide	blue	copper(II) hydroxide
iron(II) Fe^{2+}	sodium hydroxide	dark green	iron(II) hydroxide
iron(III) Fe^{3+}	sodium hydroxide	rusty brown	iron(III) hydroxide
lead(II) Pb^{2+}	potassium iodide	bright yellow	lead(II) iodide

D3

How does flame testing work?

Flame colours are produced by the movement of electrons in a metal ion. When a metal ion is heated it gains energy. Electrons in the ion become excited as they are boosted to higher energy levels. As they fall back, light is emitted. The colour of the light depends on the electron structure of the ion.

What happens in a precipitation reaction?

Precipitates form when ions in solution combine to form an insoluble salt. We can show this using an ionic equation:

$$Ag^+ \quad + \quad Cl^- \quad \rightarrow \quad AgCl$$

silver ions (e.g. in silver nitrate solution) + chloride ions (e.g. in sample solution) → silver chloride (solid)

Discuss

In November 2006, Russian dissident Alexander Litvinenko was poisoned with radioactive polonium-210 during a meal in a restaurant in London. Some toxic heavy metals and radioactive materials are odourless, tasteless, and lethal in tiny quantities. Discuss the ways in which toxicologists can detect these, before and after death.

🔍 ...flame testing colours

Detecting alcohol

The drunk driver

A person is seen to be driving erratically. The car swerves onto the pavement and knocks down the brick wall of someone's garden. Onlookers suspect that the driver has been drinking alcohol.

Several hours later, a woman is found asleep in her car a few miles away. The car fits the description and has a large dent and traces of brick dust at the front. The police officer breathalyses the woman.

P4

Detecting alcohol

When a police officer stops a driver suspected of being over the **legal alcohol limit**, he asks the person to blow hard into the breathalyser device. This means that an air sample from deep inside the lungs will be tested. The device works on a traffic light system. If the green light shows, the person is below the legal limit. If the amber light comes on, the person must be taken back to the police station for further tests. If the red light lights up, the person is definitely over the legal limit.

After an amber result with a breathalyser, the police need to get an accurate measurement of alcohol in the driver's body. Back at the police station, the person is asked to blow into a machine called the **Intoxilyzer**. These machines are calibrated using a solution with a known concentration of alcohol.

If the Intoxilyzer gives a borderline result for the level of alcohol in the breath, an additional test is required. The police doctor takes a sample of the driver's blood and analyses this using gas chromatography.

The modern breathalyser is called the Lion Alcolmeter®. It uses a fuel cell sensor to measure alcohol

Know more

The legal limit in the UK for alcohol in a person's breath is 35 µg of alcohol per 100 cm^3 of breath.

🔍 ...Intoxilyzer

M3

Know more

The first type of breathalyser was used in 1967.

The suspected drink-driver was asked to blow through a tube into a bag of potassium dichromate crystals. In the presence of alcohol, the crystals turned from orange to green. The more alcohol there was in a person's breath, the more crystals turned green.

How the Intoxilyzer works

The Intoxilyzer uses a technique called infrared spectroscopy. This identifies chemicals by analysing the chemical bonds present in the molecule. Different types of bond absorb different wavelengths of infrared radiation.

The chemical bonds in the alcohol molecules absorb infrared radiation passing through a chamber in the Intoxilyzer.

The more alcohol present in the sample, the greater the absorption. When the person breathes into the machine, the concentration of alcohol in the person's breath is measured and calculated.

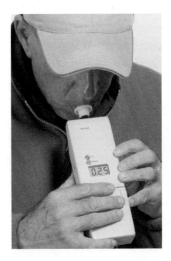

D3

What happens when we have a drink of alcohol?

Shortly after having an alcoholic drink, the alcohol is absorbed into the bloodstream. It can be detected in a person's breath, and when the body starts to excrete alcohol it also appears in the urine. Legal limits have been set for the concentration of alcohol in the blood and the urine, as well as in the breath.

The difficulty with all the tests is that they measure alcohol level at the police station, when the incident may have occurred many hours before. Even if results are negative or borderline, it is possible that the driver was over the legal limit when the incident took place.

The rate at which people break down alcohol in the body varies from person to person. A person might remove from 10 to 25 mg of alcohol per 100 cm^3 of their blood in an hour. The forensic scientist has to do a **back-calculation** to estimate the concentration of alcohol in the person's blood at the time of the incident.

Think about

The legal limit in the UK for concentration of alcohol in the blood is 80 mg per 100 cm^3. If a person is found to have a level of alcohol of 70 mg per 100 cm^3 of blood two hours after being apprehended for erratic driving, what could be concluded?

The forensic scientist's report

Your assessment criteria:

P5 Outline the results of the investigation as a statement to the court

M4 Explain the conclusions drawn from the investigation as a statement to the court

D4 Justify the conclusions drawn from the investigation as a statement to the court

The case for the prosecution

Julia is a forensic scientist who works for the Forensic Science Service (FSS). She writes up the results of her forensic examination of the evidence as case notes. Julia will use these case notes to write up her report, which will be used by the prosecution lawyers in court.

P5 The forensic scientist's report

'My report needs to summarise the investigative work I've carried out, and what I've found out. It has to have all the necessary detail. It also has to be written in a way that makes it easily understandable to police, lawyers, the jury or a magistrate, as none of these will be scientists.

'I include as much detail in my report as possible. I try to think of the questions the lawyers might ask in court. If I've answered all these questions, it's possible that I won't be called to give evidence in court.'

M4 Explaining the findings

The forensic scientist must explain the results of her findings. Sometimes more than one interpretation is possible. It's essential that this is pointed out in the report. She can give an opinion, as an expert, as to which interpretation is the most likely.

D4 Commenting on the findings

Any comments that the forensic scientist makes on the results must be purely science-based and relevant to the investigation. For instance, if a DNA analysis has been carried out, the forensic scientist can give the *probability* of the accused not being at the scene of crime. Again, based on scientific knowledge and understanding, she could comment on whether she thinks the level of a drug in a person's body contributed to their death.

Think about

Although the evidence may be being used for the prosecution, it's possible that some of the findings could favour the accused. Can you think of an example of evidence that may make this possible?

...forensics in court

Professional ethics are important. The forensic scientist must be impartial and objective. She has to disclose all findings of her investigations. She can't withhold any of the information that does not support the prosecution's case. It all has to go into the report.

D4

Section of report	Contents of report
Details of forensic scientist	My report always begins with my full name, address of the lab and my qualifications as a forensic scientist
Case against	This gives the full name and date of birth of the person accused of the crime
Lab reference number	This is the unique number issued to the case
Health and Safety issues	I have to give a warning to those handling the evidence in court if there are any hazards and risks associated with handling the evidence
Background to the case	I provide some background to the case so that readers of the report can put the various tests that I've carried out into context
Receipt of exhibits	This is a list of the evidence I received and examined. It's not always necessary to examine all the evidence. I make a note of the number of items that I didn't examine. I also include information about the transfer of these items
Technical details	This section includes an explanation of what each piece of evidence was being tested for, and a description of each test carried out. The tests I used must be accepted scientific standard procedures. I also include details of any scientific instruments used, and some background to the techniques, e.g. DNA fingerprinting. I also include any technical issues that I've encountered
A summary of my results	I include all the results of my testing, including negative as well as positive findings. All results must be recorded consistently, and include appropriate SI units. Calculations and statistics must be shown
Interpretation	I must include an interpretation of the results, and I can comment on the results
Additional staff	I have to give the names and qualifications of my assistants, who may have carried out some or many of the tests
Statement of truth and signature	I provide a statement of truth which confirms that I believe the contents of the report are true
Signature	I am responsible for the report, and must sign to confirm this. The report can't go to court without my signature

🔍 ...forensic science code of ethics

Forensic science and the law

Your assessment criteria:

P6 Discuss the role of the Forensic Science Service in the criminal justice system

M5 Identify the links between the Forensic Science Service and the criminal justice system

D5 Explain the relationship between the Forensic Science Service and the criminal justice system

My day in court

Forensic scientist Julia has worked closely with the police in the investigation of a crime. She has analysed all the different types of evidence sent to her and produced her report. The severity of the crime means that she will now have to go to court and present her evidence in person.

P6 Forensic science and the criminal justice system

The **criminal justice system (CJS)** is one of the major public services in the country. It is responsible for detecting crime and bringing it to justice.

The CJS includes

- the Police Service
- the **Crown Prosecution Service (CPS)**
- Her Majesty's Court Service
- the National Offender Management Service (prisons and probation)
- and the Youth Justice Board.

The purpose of the CJS is *to deliver justice for all*. This includes punishing the guilty and helping them to stop offending, while protecting innocent people.

The **Forensic Science Service (FSS)** is the biggest provider of forensic science to the CJS in England and Wales. The Police Service, the CPS and Her Majesty's Court Service link closely with the FSS in bringing a person to justice.

Know more

Since 2005, the FSS has been owned by the government.

...England court system

Know more

Only people who contest their guilt are actually tried. Many people, when confronted with the evidence that's been compiled against them, will plead guilty to some or all of the charges.

Know more

A jury is made up of 12 members of the public. Individuals are selected at random. Anyone who has been a resident of the UK for a period of at least five years and is between the ages of 18 and 70 can be called to give jury service.

The case goes to trial

When there is sufficient evidence that someone has committed a crime, a prosecution can be brought, usually by the CPS.

In England and Wales, criminal justice is based on **adversarial justice**. Prosecution and defence lawyers each present their cases. The guilt of a person is then decided on the strength of these cases.

The person standing trial is called the **defendant**. It is down to the prosecution to prove the guilt of the defendant. The defendant is presumed innocent until proven guilty *beyond reasonable doubt*.

Sentencing

Magistrates and **judges** need to take into account the facts of the case and the circumstances of the offender when deciding the sentence to impose. A sentence needs to give the offender a fair and appropriate punishment.

The court system in England and Wales

There are two types of criminal court. Which one a case is brought to depends mostly on the seriousness of the crime.

- The **Magistrate's Court** is where less serious crimes are tried. Judgement of a person's guilt is made by the magistrate. Around 95 per cent of criminal trials start in the Magistrate's Court, even if some are later referred to the Crown Court.

- The **Crown Court** is the higher of the two courts. The most serious crimes, such as murder, rape, and death by dangerous driving are tried in the Crown Court. Individuals are tried by **jury**.

Pre-trial disclosure of all evidence

M5

Anita's case is now being brought to trial at the Crown Court. She describes the process.

*'All documentary evidence relating to the case, including my report, must be available to be looked at, both by the prosecution **and** the defence. This is called pre-trial disclosure.'*

The defence lawyers often engage a forensic scientist to review the evidence against the defendant. This ensures that the defendant is given a fair trial. Forensic scientists employed by the defence are usually from small independent forensic science companies, rather than the FSS. They don't usually carry out their own scientific tests on evidence.

Examination and cross-examination

'I'm allowed to take a clean copy of my report into the witness box. I'm not allowed to annotate it.

I give my evidence under oath. I'm then asked to give my name, qualifications and expertise.

'I'm first examined by the prosecution lawyer. I'm usually asked questions that take me through the findings of my report.

'Then I'm cross-examined by the defence lawyer, who could try to challenge the accuracy of my test results. If there was any possibility of cross-contamination of evidence occurring, my findings would be thrown into doubt. My interpretation of the results may also be questioned. The defence lawyer may try to persuade the court that other interpretations of the evidence are possible. I have to defend my conclusions but also acknowledge that other explanations may be possible, if that's the case. I must always stick to the science involved and not make any statements that could imply the guilt of the defendant.'

Questioning witnesses

Some of the lawyers' questions to witnesses will be open. The lawyer will want a witness to describe something in their own words. **Open questions** can't just be answered by 'yes' or 'no'. Lawyers sometimes use open questions so that witnesses can create an impression on the judge and jury.

Closed questions focus on a specific point. A closed question could be, 'What type of tool was used to open the window?' Lawyers may use a closed question if the witness hasn't given specific information in their answer to an open question. But the question can't be too leading – in other words, it is not permitted to 'put words in the witness's mouth'.

Types of evidence presented in court

- **Real evidence** consists of exhibits from the crime scene and the forensic scientist's analyses.

- **Documentary evidence** will include statements from witnesses and CCTV video footage.

- **Witness evidence** given in court involves witnesses giving factual information about what they saw or heard. Ordinary witnesses are not entitled to give opinions in court.

Think about

What examples of evidence would defence forensic scientists not be able analyse, even if they thought it would be useful to do so?

Think about

Explain how the clear chain of handling of items, and the careful training of SOCOs, lowers any risk of contamination of evidence.

Think about

Think of examples, in crime scene situations, where open and closed questions could be used.

...defence lawyer

M5

CCTV footage is documentary evidence

- **Professional witnesses** include people such as police officers who provide factual evidence in court. They are allowed to contribute some opinion.

- **Expert witnesses** include the forensic scientist. The forensic scientist will be asked to give opinions as well as factual information.

D5

A sentence to suit the crime

Sentencing will depend on the severity of the crime. It must also take into account possible **rehabilitation** of the offender. Rehabilitation involves providing the offender with appropriate mental and physical experiences, so that their behaviour is changed permanently, making them less likely to re-offend.

All offences have a maximum penalty, and some have minimum sentences. A life sentence, for instance, *must* be given for murder. There is also an automatic life sentence for a second sexual or violent offence. Minimum sentences are also given for someone dealing in weapons or ammunition, and for those trafficking Class A drugs or convicted of burglary for a third time.

Community sentencing is an alternative to a prison sentence. Offenders are supervised within the community so that they are less likely to re-offend. Curfews may be imposed, and the offender may be excluded from various activities and places. When the crime committed is related to the use of drugs or alcohol, or to the person's mental health, treatment may be part of the sentence, if the offender agrees to that.

The Probation Service supervises offenders serving their sentence in the community. It also prepares prisoners for release and works with them after their release.

Appeal

Following the trial, the defence may appeal against a conviction or sentence, or the prosecution may argue for a more severe sentence in the **Court of Appeal**.

Forensic scientists often have a major part to play in appeals.

...Court of Appeal function

Assessment and Grading criteria

To achieve a pass grade, my portfolio of evidence must show that I can:

Assessment Criteria	Description	✓
P1	Carry out an investigation to collect evidence from a crime scene	
P2	Demonstrate the most appropriate methods to record and preserve evidence from the crime scene	
P3	Produce a simple plan to analyse biological, chemical and physical evidence from the crime scene	
P4	Carry out experiments to analyse biological, chemical and physical evidence from the crime scene	
P5	Outline the results of the investigation as a statement to the court	
P6	Discuss the role of the Forensic Science Service in the criminal justice system.	

Forensic scientists have to wear protective clothing to avoid contamination of evidence and risks to their own health

To achieve a merit grade, my portfolio of evidence must show that I can:

Assessment Criteria	Description	✓
M1	Describe the processing of a crime scene, explaining how the techniques used obtained valid forensic evidence	
M2	Produce a detailed plan to analyse biological, physical and chemical evidence from the crime scene	
M3	Describe the patterns found from the evidence and make connections	
M4	Explain the conclusions drawn from the investigation as a statement to the court	
M5	Identify the links between the Forensic Science Service and the criminal justice system.	

To achieve a distinction grade, my portfolio of evidence must show that I can:

Assessment Criteria	Description	✓
D1	Evaluate the processing of a crime scene, interpreting how the valid evidence collected could be used in a criminal investigation	
D2	Assess the potential risks associated with analysing biological, physical and chemical evidence from a crime scene	
D3	Explain the patterns found from the evidence and make connections	
D4	Justify the conclusions drawn from the investigation as a statement to the court	
D5	Explain the relationship between the Forensic Science Service and the criminal justice system.	

Unit 14 **Science in Medicine**

LO

Be able to investigate the range of scientific procedures used in diagnosing illness

- The first x-ray of part of a human body was taken in 1895

- Your body temperature varies throughout the day and at night it reaches its lowest point

- Diabetics need to test their blood sugar levels regularly to ensure that they do not have too much or too little sugar in their blood

- Common colds are caused by a virus, not by getting cold; some people believe we get colds more often in winter because we spend more time indoors with other people who may be ill

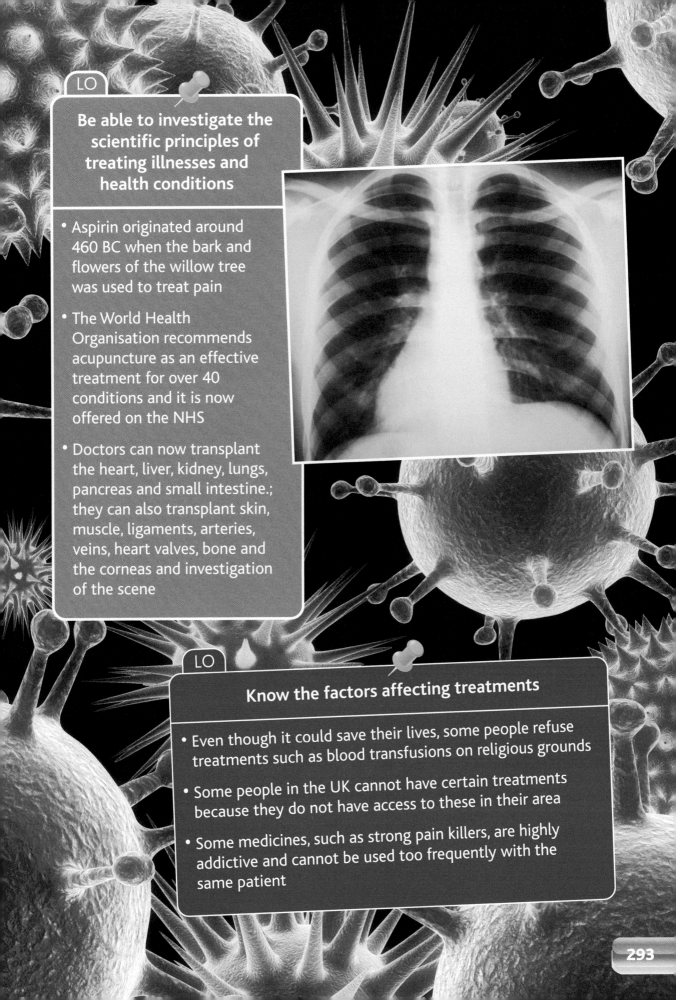

Be able to investigate the scientific principles of treating illnesses and health conditions

- Aspirin originated around 460 BC when the bark and flowers of the willow tree was used to treat pain

- The World Health Organisation recommends acupuncture as an effective treatment for over 40 conditions and it is now offered on the NHS

- Doctors can now transplant the heart, liver, kidney, lungs, pancreas and small intestine.; they can also transplant skin, muscle, ligaments, arteries, veins, heart valves, bone and the corneas and investigation of the scene

Know the factors affecting treatments

- Even though it could save their lives, some people refuse treatments such as blood transfusions on religious grounds

- Some people in the UK cannot have certain treatments because they do not have access to these in their area

- Some medicines, such as strong pain killers, are highly addictive and cannot be used too frequently with the same patient

Symptoms

Feeling under the weather

Lucy was feeling fine yesterday but in the morning she tells her mum that she is not feeling well. Her mum is suspicious. Is Lucy trying to get out of school for the day? Is she well enough to go to school or does she need to go to the doctor?

P1

Asking questions

Since Lucy is complaining of feeling ill, she will have some **symptoms**. A symptom is a visible sign or a change in the way the patient feels, or how their body is working, related to their illness. Symptoms help **diagnose** what is wrong. Examples of symptoms are having a rash or feeling sick.

For an illness to be diagnosed, the symptoms need to be assessed. We can think of them as clues. The first step is to ask the person questions about how they feel. Lucy's symptoms will form part of her **case history** – this is a detailed account of the patient's health, past health history and any other facts that might be useful in the diagnosis.

Sometimes symptoms are obvious, such as vomiting or diarrhoea. Other symptoms are not so obvious, such as feeling tired or 'run down', or having a little pain that will not go away. Often diagnosis requires clever questioning and some medical tests.

Know more

It is important to establish a time frame of the condition by clearly recording significant facts and events in the patient's medical history.

...diagnosing symptoms

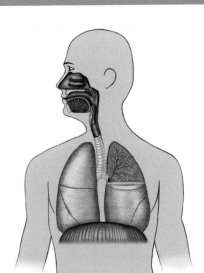

The common cold virus affects your respiratory system

What is causing the illness?

Illnesses and diseases can be caused by **biological factors** or **physical factors** (or sometimes both).

- In this case, where Lucy is complaining of feeling ill, her body has probably been invaded by a harmful organism called a **pathogen**. These organisms may be **bacteria** or **viruses**, and they affect the body's systems. The body tries to fight them off using the **immune system**. Both the invasion (**infection**) and the fight inside us cause us to feel ill, producing the symptoms that help diagnose what is wrong with us.

- Sometimes a physical problem, for example with a muscle or a bone, gives us pain and makes us feel unwell.

Discuss

Should a diagnosis of Lucy's condition be based just on what her mum has found out? Do you think parents should take a child to the doctor at the first sign of illness?

Methods of diagnosis

Doctors can assess illnesses by a **physical diagnosis**. This will include observations of physical symptoms – examining an area, checking for rashes, swellings, cuts and bruises – and possibly moving the affected area to see if there is stiffness or pain.

If a physical examination is not enough to diagnose the problem, doctors can use **biological diagnosis**. This will involve tests to find out what is happening inside the body. Samples of the patient's blood, urine or faeces may be taken for testing.

When an illness such as 'swine flu' is affecting a large number of people, it is not practical for everyone with symptoms to have individual questioning and testing by doctors. A person's condition can be assessed by call centres such as NHS Direct, or self-diagnosis using web-based information. This is quicker for the patient and cheaper for the NHS. Some critics of this scheme, however, say that there is a risk of **misdiagnosis**.

Think about

Can you think of any other advantages of diagnosis without visiting the doctor?

...NHS Direct

Taking the temperature

Feeling hot?

Tina's mum put her hand to Tina's forehead. She feels very hot, hotter than normal, and this worries her. She decides to check her temperature using a thermometer and finds out that she has 'a temperature'.

P1 Taking the temperature with a thermometer

'Taking someone's temperature' is one method of physical diagnosis. It means to measure their body temperature. A medical thermometer is used, which measures temperature on a scale of degrees Celsius (°C) or, previously, degrees Fahrenheit (°F).

The usual place for the temperature to be taken is in the mouth under the tongue. Sometimes this is not practical and it can be taken under the arm in the armpit, in the ear canal or in the anus. The temperature taken on the surface of the skin is not reliable.

A digital thermometer

Know more

*A 'normal' body temperature can range from 36.5°C to 37.2°C (97.7°F to 99°F). 'Having a temperature' means having a higher than normal temperature. A temperature of 38°C (100.4°F) or above is classed as a **fever**.*

...body temperature and fever

M1

Think about

Why do we get red in the face when we are hot, and pale when we are cold?

Discuss

What are the disadvantages of basing a diagnosis solely on the temperature of a person?

Why do we get a high temperature?

Our bodies work best at a constant (set) temperature of around 37°C. If we get too hot or too cold it can be dangerous. Our body therefore tries to control its internal temperature.

- When we get too hot our body tries to cool us down by sweating and by sending more blood to the surface of the skin (**vasodilation**).

- When we get too cold our body tries to warm us up by vibrating the muscles to generate heat (shivering), by raising the hairs on the surface of the skin (causing goosebumps), and by drawing blood away from the surface of the skin (**vasoconstriction**).

When we have an infection our body may respond by raising our body temperature. This is thought to help fight the pathogens. So a higher than normal body temperature may be a sign of an infection.

A fever is a symptom of a number of conditions

Forehead thermometers are not the most reliable

Research

Find out, compare and contrast the ways that two different medical thermometers work. Which is more accurate and why?

Evaluating thermometers

D1

Different thermometers work in different ways and can be used in different parts of the body. Older medical (or 'clinical') thermometers were made of glass and had a liquid in them that expanded when heated and rose up the glass tube. Now digital thermometers are widely used. These are safer, especially for use with children, and more accurate.

Skin thermometers are quick to use and easy to read but do not give a reliable measure of internal body temperature. When we take someone's temperature we are trying to assess the internal temperature as this is the one that should remain constant. The temperature on the surface of the skin is likely to change with the external temperature.

$\mathcal{P}$ ...medical thermometers

Checking the blood pressure

What is normal?

The doctor cannot tell what is wrong with Lucy just by taking her temperature so he is going to do some additional tests. He decides a good starting point is to check Lucy's blood pressure. He places a cuff around her arm and Lucy begins to feel a tight squeezing of her arm.

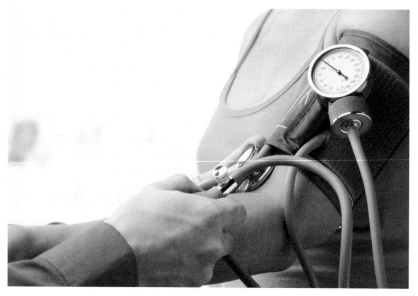

A blood pressure check

> **Know more**
>
> A blood pressure monitor is called a **sphygmomanometer**.

P1 Taking the blood pressure

Your heart pumps blood around your body to deliver the substances it needs, such as nutrients and oxygen, and to take away waste substances, such as carbon dioxide and urea. As the blood moves, it pushes against the sides of the blood vessels. The force of the blood pushing against the walls of the blood vessels is your **blood pressure**.

Checking someone's blood pressure is another method of physical diagnosis. When you have your blood pressure measured, the nurse or doctor will take two readings:

- the first is the highest pressure reached in the arteries in your arm as your heart pumps

- the second reading is the lowest pressure dropped to before the heart pumps again.

> **Know more**
>
> Blood pressure is measured in 'millimetres of mercury' (mmHg). Normal blood pressure is 120/80 mmHg, which is said as '120 over 80'.

 ...blood pressure

M1

What does blood pressure tell us?

Your heart is made of **cardiac muscle**, a kind of muscle that contracts and relaxes rhythmically all the time. Each time it contracts, it squeezes the blood inside the heart, increasing its pressure and squirting it out.

In a blood pressure reading such as 120/80 mmHg, the first (or top) number (120) is called the **systolic blood pressure**. It is the highest point that your blood pressure reaches when your heart beats. The second (or bottom) number (80) is the **diastolic blood pressure**. It is the lowest point that your blood pressure reaches as your heart relaxes between beats.

High blood pressure, called **hypertension**, increases your risk of health problems. It is not usually something that you feel. It does not tend to produce obvious symptoms. The only way to know the level of your blood pressure is to have it measured.

It is normal for blood pressure to vary within a range. Factors such as stress, exercise and certain drugs can affect it. However if your blood pressure goes above 140/90 mmHg then you should take medical advice to try to lower it.

Unusually low blood pressure, known as **hypotension**, may also be a problem. Some people with low blood pressure (perhaps 90/60 mmHg) experience symptoms such as faintness, dizziness and feeling sick. There may be an underlying cause which could need treatment.

People with hypotension have to regularly monitor their blood pressure

Think about

Why would it be incorrect to diagnose high blood pressure based on one test?

Think about

*Why does a manual sphygmomanometer need to be used together with a **stethoscope** to listen to the heart beat? Think about the pressure as the heart beats and relaxes.*

Research

Find out what kind of illnesses can be caused by persistent hypertension.

D1

Evaluating blood pressure monitoring

Different sphygmomanometers work in different ways but follow the same principles. Digital sphygmomanometers are thought to be the most accurate although manual sphygmomanometers like the one shown opposite are still popular with doctors.

There is also debate as to how frequently a person should have their blood pressure monitored. One suggestion is that a healthy person should have it checked every 2 years, and a person with a history of hypertension at least once a year.

...hypertension

Looking inside the body

That hurts!

Sara has fallen off her bike. Her ankle is badly swollen and her shoulder is very painful. She is taken to hospital where the doctor feels the areas that hurt and tries to move them. The doctor decides to arrange for some scans to help diagnose what is wrong.

P1 Body scans

Body scans are methods of seeing inside the body. They can help diagnose problems with bones, tissues and organs. There are different types of scanning method. Three commonly used ones are:

- **X-ray photography**. X-rays are a type of electromagnetic wave. They pass easily through flesh but not through bone. They can produce an image when they hit a special detector such as a photographic plate. As the x-rays don't pass through bone, a 'shadow' or outline of the bone is visible in the image. This is what the doctor uses to diagnose fractures and breaks.

- **Computerised Tomography (CT) scans**. These also use x-rays but in a different way. In a CT scanning machine, the x-ray beam moves all around the patient, scanning from hundreds of different angles. A computer takes all this information and puts together a 3D image of the body. This gives far more detail and doctors can even examine the body 'slice by slice'.

- **Magnetic Resonance Imaging (MRI) scans**. An MRI scanner is a big machine that can produce very detailed pictures of parts of the body that would not be distinguished by x-rays. It uses a strong magnetic field and radio waves to create images on a computer of tissues, organs and other structures inside your body. These can be 2D or 3D.

- **Endoscopy**. This is a technique by which a camera on the end of a small flexible rod is inserted into the body through an opening (such as the nose or mouth). It is a very useful way to diagnose obvious problems without exploratory surgery.

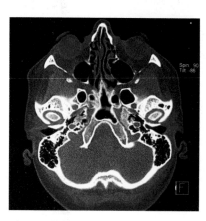

This CT scan provides a detailed picture of the head bones

> ### Know more
>
> *X-rays and radio waves are both types of electromagnetic radiation and form part of the electromagnetic spectrum. The existence of the electromagnetic spectrum wasn't just 'discovered' but was **hypothesised**, based on the work of several different scientists over hundreds of years.*

 ...MRI

M1

Discuss

What special precautions do people who work with x-rays have to take, and why?

Know more

Remember you learnt about the electromagnetic spectrum in Unit 2.

Energy and the electromagnetic spectrum

The electromagnetic spectrum is a collection of electromagnetic waves that have different properties:

radio waves, **microwaves**, **infrared**, **visible light**, **ultra-violet (UV)**, **x-rays**, **gamma rays**

The lower the frequency of the wave (or the longer the wavelength), the less energy it has and the less harm it can do to the human body.

Radio waves have the lowest frequencies (and longest wavelengths). Gamma rays have the highest frequencies (and shortest wavelengths). Gamma rays and x-rays are potentially dangerous to the human body when their energy is absorbed. But, used with precautions, they provide invaluable imaging methods for diagnosis.

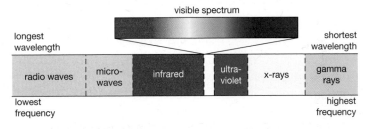

The electromagnetic spectrum

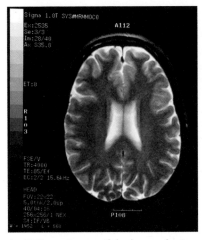

An MRI scan image of the human brain is complex

Evaluating imaging techniques

D1

Each method of seeing inside the body has advantages and disadvantages. x-ray photography is widely used and relatively cheap to carry out. Most hospitals have this type of imaging technology. If the process is digital, the results can be quick and easy to obtain.

CT scans and MRI scans are more specialised and require expensive equipment. Not all hospitals have these scanners and so there is limited availability resulting in waiting lists. The specialist diagnosis can also take some time.

It is very important that the medical practitioners taking and assessing the image know exactly what they are looking at, and what would show a deviation from normal. Training and experience are needed. People who use scans to diagnose conditions are called **radiologists**.

...CT scan

Biological diagnosis

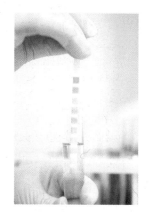

Getting to the bottom of it

Rita has been feeling unwell for several days. She lacks energy, has very little appetite, gets stomach cramps and occasionally feels dizzy. She decides to go to the doctor. The doctor suspects she has got 'a virus' but needs to be sure, so sends off some of her blood, urine and faeces for analysis.

Analysing urine

Your assessment criteria:

P1 Carry out investigations into biological and physical procedures used to diagnose illness

Practical

M1 Explain the scientific principles underlying the biological and physical procedures used to diagnose illness

D1 Evaluate the advantages and disadvantages of using biological and physical procedures to diagnose illness

P1 Types of biological test

- A **blood test** is a very useful test that will look at the composition of the blood. Checks can be made to see if the different types of blood cells are present in the correct amounts, to see if the levels of expected chemicals are correct, and to identify unexpected chemicals. A blood test can diagnose many conditions, for example **anaemia**.

- A **urine test** can tell us important things about the body as it will show what the body is getting rid of (excreting). It can diagnose **diabetes**.

- A **faeces test** is a useful test to see how well the digestive system is working and if there are any problems. It can diagnose a serious condition called colon cancer.

- **Cytological tests** (for example a cervical smear test) involve taking a sample of potentially affected tissue and observing the cells under a microscope. It can diagnose cells that might be malfunctioning or showing signs of being abnormal.

- **Genetic investigations** are relatively new tests that check if there is any problem with the person's DNA, and whether it might be or has been passed on to children. Combined with family history this is a powerful diagnosis tool that can help parents make decisions about having a baby, or diagnose conditions early in a baby's life. One genetic condition is **cystic fibrosis**.

There are many other tests, such as sputum tests. The analysis of the results of biological tests is called **pathology**. This analysis will form part of the biological diagnosis of a particular set of symptoms.

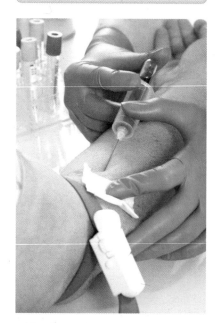

A blood test involves taking a sample of blood from a vessel in your arm. This is then sent away for analysis

Know more

The average adult human has 5 litres of blood, and each litre contains around 5 trillion (5 million million) red blood cells.

🔍 ...blood composition

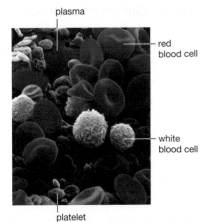

The main constituents of blood

Testing blood sugar can be done easily with a drop of blood only

Think about

When would be the right time to take samples and send them off for analysis – as soon as the patient first visits the doctor, or after other tests have been carried out?

Research

Some biological diagnosis is now routine for certain groups of people. Research the kinds of tests that are routinely done and what diseases they help to detect.

What can be detected in blood, urine and faeces? **M1**

The blood plasma carries dissolved substances, and if there is an unexpected substance it will show up in a blood test. This is useful for detecting drugs, alcohol and certain poisons. It can also show how much oxygen or carbon dioxide (or monoxide) is in the blood, and how much cholesterol there is. Too much cholesterol can contribute to heart disease.

Blood is essentially a super-highway to every part of your body. It can sometimes carry harmful 'foreign' organisms (**pathogens**), which can cause infections.

Each type of blood cell has a specialised function, so blood tests can check that these are working as they should.

- Red blood cells take up oxygen from the lungs and deliver it to the body tissues. If there are too few of them or they do not work well, we can get ill with anaemia.

- White blood cells help to fight infections. If there are too few of them or they are not working well, we are likely to get ill more frequently. This occurs in types of **leukaemia**.

- Platelets are responsible for forming blood clots, so if there are too few of them this can affect the healing of wounds.

Urine is a waste product from the kidneys. Any chemicals that need to be got rid of from the blood, such as excess glucose, will be excreted in the urine. A urine test is important in detecting these chemicals and shows whether the kidneys are working well. It enables diagnosis of conditions such as diabetes.

Faeces is the waste produced by digesting our food. Its colour and texture depend on the diet of the person. Anything that goes wrong in the intestines (for example an infection, a tumour, or parasites) can affect the faeces. A faeces test is therefore useful for detecting a range of problems, from infections that cause diarrhoea to cancer.

Evaluating biological diagnosis **D1**

Each diagnostic test has it own procedures. Some are relatively simple and can even be carried out at home, for example testing the level of blood sugar. Other tests are more difficult to carry out and require a visit to the doctor or to hospital. Few tests give instant results. Most need complex analysis that takes time and specialised trained medical personnel. This can be costly to the health service.

...blood tests

Microbes and disease

Invasion!

When we describe being ill we might use terms like 'I've caught a bug' or 'I've got flu' or 'My stomach's upset'. But what do these terms actually mean and what actually makes us ill?

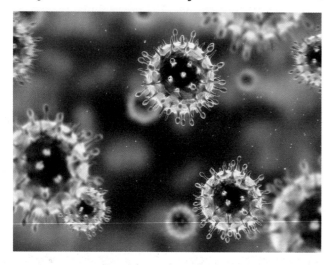

'Bugs' that make us ill

P1 — What are germs and bugs?

A germ or a 'bug' that makes us ill is a type of **micro-organism**, or **microbe**. This is a living thing that is too small to see without microscopes. Because microbes are alive, they need nutrition, they grow and they reproduce. Some microbes are useful to us but others (pathogens) can make us ill. There are three main types of pathogen: bacteria, fungi and viruses.

The illnesses caused by these can be transmitted from one person to another. The person is said to be **infectious**.

The symptoms they cause depend on the part of the body they affect. For example, a cold virus affects the lungs, throat and nose and we know the symptoms well.

Sometimes a doctor can diagnose an infection by physical means, such as looking at the person's throat or taking their temperature. If the cause of symptoms is not so obvious, the doctor can make a biological diagnosis by taking samples (for example, of sputum) and requesting tests. A pathologist will probably look at microbes in the sample under a microscope, and possibly encourage them to grow in the laboratory. The type of infection can then be identified and a suitable treatment prescribed.

Know more

Not all micro-organisms are bad. Yeast, an example of a fungus, is used to make bread dough rise.

...diagnosing viral infections

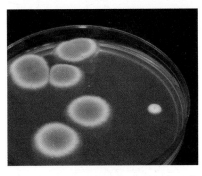

Here a micro-organism has reproduced into a colony in a Petri dish

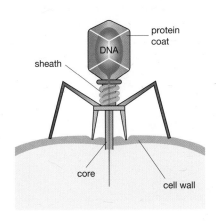

A virus attacking a body cell

Testing for microbes

M1

To test for a **microbial infection** we need to understand more about the different types of microbe and how they make us ill.

Microbe	Test
A bacterium is a **single-celled organism**, smaller than body cells. Once inside the body it can spread throughout the body (it is invasive). It makes us feel ill by producing poisonous substances called **toxins**, and by our body reacting against these.	To diagnose a **bacterial infection**, bacteria from a sample can be grown in a Petri dish in special conditions. Here they replicate and form a visible colony. Each type of bacterium forms a colony of different size and colour. They can also be viewed under powerful microscopes.
A fungus may be single-celled or may have many cells. It feeds off decaying matter. It generally make us unwell by breaking down dead or dying tissue, such as skin, and by our body reacting against this.	To diagnose a **fungal infection**, a sample can be viewed under a microscope or grown in a Petri dish in appropriate conditions to make an identifiable colony.
A **virus** is not a cell. Viruses cannot grow or reproduce on their own, but invade body cells and use them to reproduce. We feel ill because the invaded cells, and the tissue or organ which they make up, cannot work properly. Each invaded cell makes more viruses so the infection can rapidly spread. Most infected cells die.	A **viral infection** is harder to diagnose. Tests usually involve looking at the physical symptoms which are the body's response to the infection.

Research

Research the types of illness caused by bacterial infections.

Think about

Explain why it might be difficult to diagnose a viral infection.

Evaluating infection diagnosis

D1

A viral infection can be very difficult to diagnose. Sometimes a viral infection can be confused with a bacterial infection. In this case a biological diagnosis is crucial as otherwise the treatment may not be effective. **Antibiotics** are drugs used to combat bacterial infections. It would be pointless treating a cold with antibiotics because a cold is caused by a virus. But if, for example, a cough is found to be caused by a bacterial throat infection it would respond to antibiotics.

 ...diagnosing bacterial infections

Therapeutic drugs

Which medicine to use?

A common sight in many homes is a medicine cabinet. Along with plasters and bandages you are likely to find a wide selection of creams, pills, powders and liquids, all designed to help treat conditions you might have.

P2

What are drugs?

Once a doctor has diagnosed an illness, he or she will need to advise on the correct treatment. A **therapeutic drug** may be prescribed. A therapeutic drug is a substance that can be used in the diagnosis, treatment or prevention of an illness. When used as a treatment, it may help with the symptoms or it may be a cure.

The doctor will need to decide the best drug for your condition, the best way to administer it (as a tablet, capsule, liquid medicine, ointment or cream, or by injection) and the correct dosage (how much, how frequently and for how long).

Some drugs can be bought 'over the counter' at pharmacies. Common examples of these are aspirin, paracetamol, ibuprofen and cold remedies containing these.

Others have very specific action and can only be prescribed by a doctor. The doctor will write out a **prescription** which is taken to a pharmacy. The **pharmacist** will dispense it (prepare it, package it and give it out) to the patient or the patient's representative.

Certain medicines can only be obtained with a prescription

Know more

When new therapeutic drugs are developed they need to be throughly tested and trialled to make sure they are safe and effective. This often takes years and a lot of money.

...history of medicines

Therapeutic drugs can be grouped or classified according to what they do.

P2

Type of drug	Used for	Example
analgesic	treating or reducing pain	paracetamol
anti-inflammatory	reducing inflammation or swelling	ibuprofen
antibiotic and **antiviral**	affecting the life cycle of bacteria and viruses respectively	penicillin
antihistamine	reducing allergic reactions	acrivastine
chemical replacement	replacing essential substances in the body	insulin
cytological **(chemotherapy)**	acting on specific cells, tissues and organs	carboplatin
antidepressant, stimulant, sedative, antipsychotic	affecting brain function and altering our behaviour	Prozac
antagonistic	opposing the effects of other drugs or of natural body chemicals	beta blockers

Drugs designed to relieve symptoms, for example an analgesic, reduce the effects of say a cold. However, they would not get rid of the cold virus.

Aspirin is a common analgesic or 'painkiller'

Discuss

Why are some drugs available over the counter but others only by prescription?

Research

Find out how anti-inflammatory drugs work in the body.

Each drug is different

D2

Drugs are chemicals. They interact with the body to either help control symptoms or to change the way the body works. If there is an infection, they can also interact with the organism causing the infection.

Once a doctor has diagnosed an illness, he or she decides on a treatment. Any therapeutic drug chosen needs to be appropriate for the specific illness. If the illness is caused by something that can be cured, the drug chosen will aim to cure it. If the illness cannot be cured, the symptoms might have to be controlled, for example pain.

Each type of chemical in drugs has a specific way of interacting with the body. Here are two examples:

- A painkiller interacts with the nerves that transmit pain, blocking their signals; if the signals cannot travel to the brain, the brain does not interpret the pain and we do not feel the pain.

- An antibiotic either acts on the cell membranes of bacteria, causing them to rupture, which kills the bacteria; or it prevents the bacteria from reproducing.

 ...types of medicines

Different drugs

Needlephobic?

Are you one of those people who can't stand needles? The good news is that there are many other ways you can receive your medicine. The bad news is that sometimes only an injection will do.

P2 Drug formulations

Once a doctor has decided that you need a medicine, he or she must decide the best way to deliver that medicine. The form a medicine takes is known as its **formulation**. It may be a tablet or capsule, a liquid medicine, a liquid that needs to be injected, a cream or ointment, or a skin patch. Each formulation is administered (given) in a particular way. For example, this may be:

- **topical** – spread, rubbed in, or sprayed on the area that needs treating

- **oral** – swallowed

- **inhaled** – breathed in through the mouth and nose

- **injected** – inserted by needle, either directly into a blood vessel (intravenous injection) or under the skin (subcutaneous injection).

This treatment for asthma is a liquid, taken via inhalation from a pump

Formulation	Administration
cream	topical
ointment	topical
patch	topical
tablet	oral
capsule	oral
oral liquid	oral
spray	topical , oral or inhalation
injection liquid	intravenous or subcutaneous injection

Examples of formulation and administration

Know more

Fear of needles is called trypanophobia.

 ...topical medications

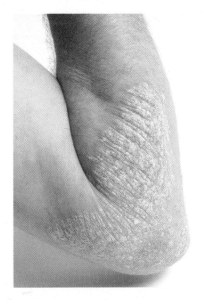

Psoriasis is usually treated by topical methods because it is a skin condition

Deciding how to administer a drug

M2

The formulation of a medicine is directly related to the type of condition it is used to treat.

Psoriasis is a very painful skin condition. The person develops a rash, intense itchiness, and cracks in the skin that may bleed. It can be a lifelong condition that comes and goes. The main treatment method is the management of the symptoms when they arise.

When psoriasis flares up, the first treatment is to sooth and moisturise the skin. Applying topical skin creams helps to keep the skin moist and supple, reducing redness, scaling, dryness and cracking. The choice of moisturiser or **emollient** will depend on the type of psoriasis and personal preference.

Other topical treatments may be prescribed. Steroid creams can help to reduce inflammation. Vitamin D creams may also be beneficial.

In severe cases of psoriasis, oral or **intravenous** treatments may be considered.

Injections

D2

Injections are used to deliver liquid medicines directly into the blood or tissues. Generally they are fast-acting. They are used to deliver drugs such as analgesics, **vaccines** and **anaesthetics.**

If a person is in a great deal of pain, for instance, morphine would be injected intravenously. Most injections need to be delivered by trained medical professionals. However, if a person has a chronic lifelong condition that needs regular or emergency medication by injection, they can be trained to deliver it themselves. Many people with diabetes inject themselves **subcutaneously** with insulin every day, to help control their blood sugar level. This is called self-administration.

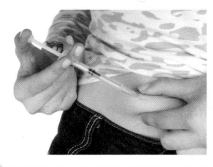

Self-administering insulin

Research

Research the different types of anaesthetic needed during operations such as sports injuries and caesareans. Find out how these anaesthetics are administered.

Research

Research 'drug implants'. What types of drugs are delivered in this way?

🔍 ...intravenous injections

Physical therapies

When drugs aren't the answer

For some people, drugs don't provide a remedy. While drugs can help with many conditions, with others they are ineffective. Some other form of treatment is needed in these cases, either on its own or together with drugs.

P3

What is a physical therapy?

A **physical therapy** is a treatment that does something physically to the body. There are many kinds of physical therapy.

Type of therapy	Used for	Example
surgery (operating)	Exploring, repairing, removing and replacing internal structures.	appendectomy
radiotherapy	Killing tumour cells.	cancer
laser therapy	Using high-energy lasers to kill or remove tissue.	laser eye surgery to cure short-sightedness
physiotherapy	Treating muscular and skeletal problems.	sporting injuries
osteopathy	Treating skeletal or muscular problems but also considering other factors such as lifestyle, diet and posture.	treating a bad back
'alternative therapies', such as **acupuncture**, **chiropractics**, **reflexology**	Treating a wide range of conditions, often as a **complementary therapy** alongside a conventional therapy.	chronic pain
replacement therapies	Replacing parts of the body that are damaged or not working well.	blood transfusion, organ transplant

Each type of therapy has a different set of ideas behind it; for example, acupuncture is based on an ancient Chinese idea of 'energy flow' in the human body.

Your assessment criteria:

P3 Carry out investigations into the scientific principles of physical therapies used to treat given conditions

> Practical

M3 Describe how physical therapies would be used to treat given conditions

D3 Assess, using scientific and other evidence, which physical therapies are effective in the treatment of conditions

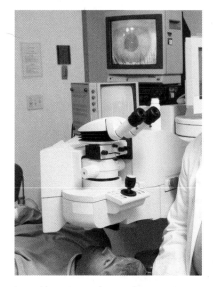

Laser therapy can be used to cure short-sightedness

Know more

Some physical therapies, such as acupuncture and reflexology, have been used for many centuries. Other physical therapies are more recent. Laser eye surgery was only developed in the 1990s.

🔍 ...complementary therapies

M3

D3

Research

Research two different approaches to treating the same condition, for example short-sightedness or back pain. In what ways are the treatments different?

How physical therapies are used

Some physical therapies seem very similar but to the practitioner they are very different. Take the difference between osteopathy and chiropractics. They both treat back pain but there is a difference in approach.

When an **osteopath** treats a bad back he or she manipulates the muscles, ligaments, tendons, tissues and organs related to the back pain. The aim is to aid the body's natural healing ability.

A **chiropractor** manipulates the bones in the back to ensure they are in alignment, have full movement and are not pressing on any nerves.

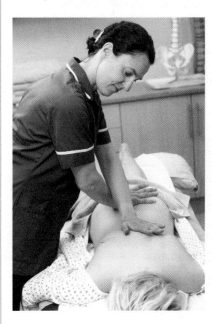

Back pain can be an extremely disabling condition. Osteopaths and chiropractors can help reduce pain and aid recovery

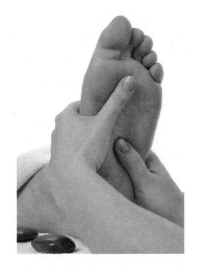

A reflexologist believes that parts of the foot correspond with other parts of the body, and that massaging and manipulating the feet can help treat many different problems

Research

Many people acclaim the benefits of acupuncture for all sorts of conditions. Find out what evidence there is that acupuncture really works.

Assessing physical therapies

There are many physical therapies to choose from. Some are based on accepted science whereas others are based more on tradition. Even if science cannot explain how a therapy works, its effectiveness can be judged from **evidence**. Evidence can range from studies, tests and data generated by scientific trials to anecdotal evidence (asking individuals if it works for them).

Recently there has been a lot of disagreement about offering complementary therapies, such as reflexology, as part of the National Health Service (NHS). While there is a lot of anecdotal evidence for their success, when the effectiveness of an alternative therapy is tested in a scientific way it often fails. People who believe in these therapies argue that it is the testing that is flawed, not the therapy.

🔍 ...osteopathy

Physiotherapy treatment

Your assessment criteria:

P3 Carry out investigations into the scientific principles of physical therapies used to treat given conditions

Practical

M3 Describe how physical therapies would be used to treat given conditions

D3 Assess, using scientific and other evidence, which physical therapies are effective in the treatment of conditions

A day in the life of a physiotherapist

Lea is a physiotherapist. She treats people with all sorts of conditions: respiratory problems, people who have had strokes or other neurological problems, people who are recovering from broken bones, and those who have had new hips, new knees, etc. No day or case is ever the same for her.

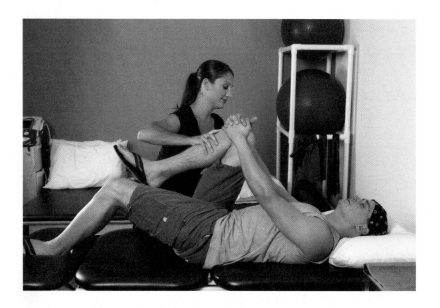

P3

Locating the problem

When a physiotherapist first sees a patient, he or she asks the patient some questions to note down a case history. For example, Lea would probably ask:

- *Where is the pain?*
- *What kind of pain is it?*
- *How long have you had it?*
- *What treatment have you already had?*
- *Are you taking any drugs?*

She also needs to find out what their lifestyle is like, what kind of job they do and if anything particular is affecting their pain. She might then massage or manipulate the area that is painful, all the time asking the patient what it feels like. If any x-rays or other diagnostic tests have been done she would review those results.

Know more

Sometimes symptoms such as pain in one specific area are obvious, but 'referral pain' may not be in the area of the injury.

 ...physiotherapy

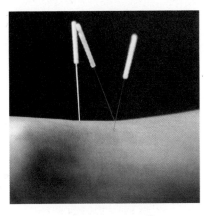

An acupuncturist inserts needles at particular points to help control swelling and pain

M3

Common conditions that a physiotherapist might see regularly are:

- sporting injuries, like pulled muscles, ripped tendons and ligaments, possibly fractures

- swollen joints, bruises, painful joints or muscles

- chronic (long-lasting) conditions such as deformed joints.

Patients may be taking analgesics or anti-inflammatory drugs to help cope with the painful symptoms of the condition.

Treatment methods

Physiotherapy treatment includes:

- Treating pain. Sometimes massage and 'hot and cold' treatments can relieve pain; in other cases the underlying cause, like a muscle spasm, needs to be dealt with; sometimes other treatments such as ultrasound can help.

- Getting things moving. Sometimes the patient can't move an area or a joint; physiotherapists work with them to encourage movement in the joint or muscle, by manipulating the area or massaging it and by advising special exercises.

Arthritis is a painful condition. Physiotherapy aims to help joints move and to control the swelling and pain

- Complementary therapies. Sometimes conventional physiotherapy alone is not enough to help a patient. Physiotherapists consider all possibilities, for example acupuncture or **aromatherapy** massage, that might complement the physical therapy.

> **Think about**
>
> *What kinds of injuries might a football team physiotherapist need to treat? What treatment might be tried?*

> **Research**
>
> *Find out what aromatherapy involves.*

Assessing physiotherapy

D3

Strong evidence of the effectiveness of physiotherapy comes from the scale of its use in the National Health Service. Whereas some therapies are not routinely used, physiotherapy is standard treatment at all hospitals for a wide range of conditions.

 ...aromatherapy

Surgery

Your assessment criteria:

P3 Carry out investigations into the scientific principles of physical therapies used to treat given conditions

  Practical

M3 Describe how physical therapies would be used to treat given conditions

D3 Assess, using scientific and other evidence, which physical therapies are effective in the treatment of conditions

Drastic action

You have probably seen the action in an operating theatre in TV programmes – the excitement, the tension, the blood, and the saving of lives. Surgery can actually range from minor, routine operations to highly specialised, complex procedures. The surgeons who carry out this type of therapy are very highly trained.

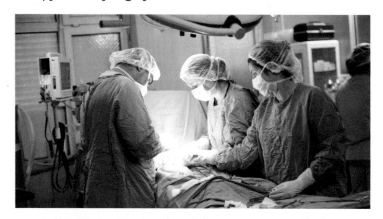

P3 Types of surgery

Surgery is an 'invasive' therapy, which means the surgeon needs to go into the body to carry out the treatment. Surgical procedures are classified as:

- day surgery – routine surgery after which the patient usually leaves hospital the same day

- elective surgery – surgery planned in advance, usually involving a short stay in hospital

- emergency surgery – operations that need to be done immediately.

There are different surgical methods. Keyhole surgery is done through a small hole in the skin. **Microsurgery** involves using very powerful magnifiers.

For some operations **local anaesthetic** is used. The area being operated on is numbed but the patient remains awake. For longer or more complex operations the patient is given **general anaesthetic** and is then asleep throughout the operation.

Some types of operation have special names, for example an *appendectomy* is the removal of an appendix, and a *rhinoplasty* is the repair or modification of the nose.

Know more

Although anaesthetics in some form (drugs like opium and alcohol) have been around for centuries, it was only in 1844 that nitrous oxide ('laughing gas') was used when pulling out teeth.

Know more

'tomy' means cutting;

'rhino' means nose and 'plasty' means repair.

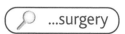
...surgery

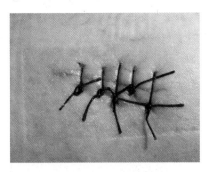

Surgery often requires internal and external stitches

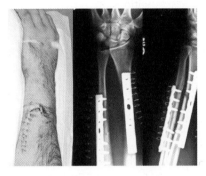

This arm had bad breaks in two bones and needed to be repaired using pins

What surgery is needed?

M3

The type of surgery that a patient needs depends on the condition being treated.

- If the illness or internal damage is unknown, then 'exploratory' surgery might be needed.

- Internal parts of the body that are damaged or not working properly might need to be repaired. For example, if a blood vessel or muscle is damaged in an accident it might need to be repaired by surgery.

- If a piece of the body is severely damaged or cancerous then it might need to be removed.

- A limb (part of an arm or leg) will only be removed (amputated) if the damage is beyond repair.

For some conditions there are options available to the surgeon. If a person has **appendicitis**, their appendix is inflamed and might become infected. This can be dangerous so the appendix needs to be removed. The surgeon can either cut into and open up the abdomen and cut the appendix out, or he can make a small **incision** and, using probes and fine instruments, remove the appendix through the small hole (this is keyhole surgery).

Other types of surgery can be **non-invasive**, for example laser eye surgery. An endoscopy tube can successfully be used to remove obstructions inside the body without cutting into the body.

If a person has a fractured arm it can be repaired in different ways. The usual way is to put the arm into a splint or plaster until the bone can repair itself. An orthopaedic consultant would need to assess the break by evaluating the evidence, such as looking at x-rays of the arm. If the break is very bad the bone might not heal properly. In this case it might require surgery to 'pin' the bone together using metal plates. Only by using scientific evidence could the consultant decide on the best treatment option.

The effectiveness of surgery

D3

Since surgery is a risky procedure, the effectiveness of major surgery is often evaluated by the 'survival rate'. This measure is influenced by the complexity of the surgery, the person's reaction to the anaesthetic, and the ability of the patient's body to recover.

...history of surgery

Replacement therapies

Your assessment criteria:

P3 Carry out investigations into the scientific principles of physical therapies used to treat given conditions

Practical

M3 Describe how physical therapies would be used to treat given conditions

D3 Assess, using scientific and other evidence, which physical therapies are effective in the treatment of conditions

When things go wrong

When things go badly wrong and our bodies cannot repair themselves or be repaired, a more serious form of therapy is needed. Replacement therapy literally 'replaces' the part that is not working with an alternative part. But there are limits to what can be replaced in a human body, and it is a very complicated procedure.

We are not mechanical robots – not all parts can be replaced

P3

What can be replaced?

In order for replacement therapy to be successful, there must be a very close match between the replacement part and the original body part. This minimises the risk that the replacement part is not accepted but rejected by the patient's body.

Parts of the human body that are most commonly replaced are:

- tissues such as skin (known as **skin grafts**)

- blood (known as a **blood transfusion**)

- organs such as kidneys, liver, heart and lungs (known as **organ transplants**).

Let's look at blood. Everyone's blood has the same basic functions, but there are very subtle yet important differences in the composition of the blood from person to person. There are different blood groups in humans. Blood from one person cannot be transfused into a person who has a non-compatible blood group; in fact it can be fatal.

Donor blood can save the lives of car crash victims

Know more

There are four blood groups: A, B, O, AB. To make matters more complicated, there is also another possible difference called the 'rhesus factor' which can be + or −. (See Unit 13). So your blood type can actually be one of eight combinations.

🔍 ...blood transfusion

Think about

There is a waiting list for organs for transplant. Organs come from people who have given their permission before they died for their organs to be removed and used on their death. How would you persuade someone to join the organ donor list?

Research

What conditions can be treated with donations from living patients?

Conditions requiring replacement therapies

M3

When would a blood transfusion be necessary? Blood transfusions can be life-saving in some situations, such as massive blood loss due to an accident. Transfusions can also be used to replace blood lost during surgery.

Blood transfusions can be used to treat many conditions:

- severe anaemia (low red blood cell count)
- **haemophilia** (where blood lacks the ability to clot)
- sickle-cell anaemia (a hereditary condition where the red blood cells are deformed and do not work well)
- leukaemia (causes low white blood cell count).

Although patients can be given whole blood, it's more likely that they are given a part of the blood; for example, white blood cells in the case of leukaemia, or plasma products (parts of the blood plasma) that can be used to treat specific conditions such as haemophilia.

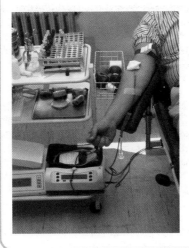

In a blood transfusion, blood or its components are transfused directly into the circulatory system of the patient

Assessing replacement therapies

D3

An organ transplant is the 'last chance' for some patients. There is no guarantee that organ transplants will work, as their success depends on a number of different factors.

Survival rates are increasing as new drugs are developed and post-surgery care becomes more effective. As technology improves, more sophisticated transplants are possible.

The chance of an improvement in quality of life, or of an extension to life expectancy, is often all that the patient and the family want. To them, even with the risks involved, transplants are an effective way of treating some conditions.

Discuss

Since organ transplant can fail, why should we bother?

 ...transplants

Preventative therapies

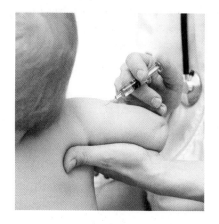

Can we prevent disease?

It is an important day for Max. He is at the doctor's surgery for his MMR vaccination. This vaccination will prevent him from getting mumps, measles and rubella, any of which could make him very ill.

Your assessment criteria:

P3 Carry out investigations into the scientific principles of physical therapies used to treat given conditions

Practical

M3 Describe how physical therapies would be used to treat given conditions

D3 Assess, using scientific and other evidence, which physical therapies are effective in the treatment of conditions

P3 · What is preventative medicine?

The idea behind **preventative therapy** is to try to intervene to prevent a disease from affecting people. Health professionals do this in three ways:

1 Prevent the disease before it happens, by vaccination.

2 Detect the disease before it takes hold, by **screening** (for example, for breast cancer).

3 Minimise the impact of the disease if someone already has it (for example, using drugs with HIV patients to reduce their chance of getting a secondary disease).

To prevent disease, medical professionals must understand who the disease is most likely to affect. Therefore the prevention is aimed at different groups of people:

- *universal*: everyone, as in case of polio vaccination

- *indicated*: those who show early signs or those who have an increased risk, such as drug users in the case of HIV

- *selective*: specific groups of people or individuals who are most vulnerable, such as the elderly in the case of flu.

Know more

The word 'prophylactic' is used to describe any procedure whose purpose is to prevent, rather than treat or cure a disease.

M3 · Conditions requiring prevention therapies

Prevention therapies are used for a wide range of conditions. Some are very general and some are very specific. For example, a vaccination only works in preventing a specific disease.

 ...vaccinations

FRANK

0800 77 66 00   talktofrank.com

'Talk to Frank' is a government campaign to educate young people about the misuse of drugs

The most common diseases for which vaccines are given in the UK are:

influenza (flu), meningitis, mumps, measles, rubella ('German measles'), tuberculosis (the BCG vaccine), polio, tetanus, diphtheria, whooping cough, cervical cancer.

New vaccinations are being developed all the time.

General health-awareness campaigns are also preventative therapies. The most common ones are:

- risks of smoking
- risks of alcohol
- risks of drugs
- benefits of physical exercise
- benefits of healthy eating.

Discuss

Do you think anti-drugs campaigns can be effective in preventing drug misuse?

Swine flu is easily transmitted; some evidence suggests that washing your hands can help to prevent the spread of flu

Research

Research how antibiotics can be used as prophylactics for some 'at risk' people.

Assessing preventative therapies

Some preventative therapies are controversial, because it is unclear how effective they are. As with all therapies, there needs to be a fair evaluation of the evidence to judge their effectiveness or **efficacy**. Debate arises because therapies need funding, because they may not be available to all, and because there may be unknown side-effects.

The new 'swine flu' vaccine is an example. The vaccination programme started on 21 October 2009 and was intended for those groups of people considered most at risk from 'complications' from the infection. The first groups eligible for vaccination were:

- pregnant women
- people who live with someone whose immune system is compromised (for example, people with cancer or HIV/AIDS)
- people aged 65 and over in the seasonal flu vaccine 'at risk' groups
- young children aged over six months and under five years.

Normal flu vaccine is between 70% and 80% effective, and it is anticipated that the swine flu vaccine will be at least that. However, there is no evidence as yet. Despite this, the health service considers it better to vaccinate than not.

...MMR

Side-effects of drugs

Your assessment criteria:

P4 Identify the general risks of specified treatments

M4 Describe, using scientific evidence, the particular risks involved in specified types of treatment

D4 Explain why some individuals may choose not to take advantage of all types of available treatments

The right one for you

Jason is a taxi driver. He has developed a bad cold and needs to buy some cough medicine. He needs to make sure he buys one that doesn't have the side-effect of making him drowsy as he could put his passengers at risk.

P4

Side-effects

Medicines are drugs that are used because of their therapeutic effects. These are the desired effects of the medicines. They can alleviate symptoms, stop the progress of the illness, or cure the illness.

Medicines can also have effects on the body that are not desirable. These are called **side-effects**. Sometimes they are serious, but most of the time they are not.

All medicines, whether purchased 'over the counter' or prescribed by your doctor, will have information about these possible side-effects. This is usually in the form of a leaflet inside the packet. Possible side-effects are listed along with other useful information. It is recommended that all patients read the leaflet before they start taking their medicine.

The risk of side-effects can affect the decision the patient makes about a drug. For example, antihistamine is very effective at controlling symptoms of hayfever, but can make people sleepy. The person would need to judge whether it was suitable to take.

A more extreme example is that of general anaesthetic, which can have serious side-effects. But it is needed when a patient undergoes complex surgery. In this case the benefits outweigh the risks for most people.

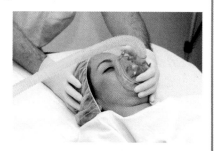

General anaesthetic can pose risks

Know more

General anaesthesia is more risky for people with certain underlying health problems such as heart disease, and for elderly people. The benefits and risks for these people need to be carefully discussed.

Know more

Some people are allergic to some medicines. If this is the case the treatment can cause severe side-effects and can even be fatal.

 ...side-effects of medicines

M4

Think about

What reasons might there be for someone affected by side-effects to continue to take a treatment?

How common are side-effects?

To explain the chance of getting a particular side-effect, information on the leaflet supplied with most medicines uses the terms *common*, *uncommon*, *rare* and *very rare*.

Description of side-effect	How many people it affects
common	1 in 10
uncommon	1 in 100
rare	1 in 1000
very rare	1 in 10 000

For all the possible side-effects, the leaflet describes what symptoms to look for.

Ibuprofen is a common painkiller and anti-inflammatory drug. Its potential side-effects can be found on the patient information leaflet in the packet. It can affect the stomach, intestines, blood, liver, kidney and nervous system.

The information about side-effects comes from the scientific trials carried out before the drug is released into the market.

The leaflet also gives advice on what to do if you have the listed symptoms. It may be important to stop taking the medicine immediately.

Addiction

Addiction to some prescribed medications (for example, sleeping pills) is a particular problem and side-effect. This means that the patient can become dependent on the medicine. The possibility of addiction is clearly listed in the patient guidance and discussed with the patient prior to taking the medicine.

According to newspaper reports, Michael Jackson had an addiction to painkillers

Reasons not to take a drug

D4

Side-effects, allergies and a possibility of addiction are good reasons for some people to choose not to take advantage of available drugs. They will consider other drugs or other therapies that do not have these effects.

Think about

Explain why some people may choose not to take a medication.

🔍 ...addiction to prescribed drugs

Choice of treatment

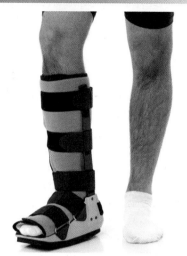

What would be best?

It is never easy when you are ill or injured. All you want to do is get better. Imagine being in a situation where you have to choose between different treatments. How do you decide? What if an effective treatment is not available where you live?

In some cases a ruptured Achilles tendon can be repaired without surgery

P5 Factors affecting the choice

There are many factors that affect the choice of treatment for any given condition.

- Medical advice: the treatment may have risks or harmful side-effects.

- Individual choice: the patient chooses (or accepts) the treatment, or declines it, for whatever reason.

- Ethical considerations: there may be reasons for believing it to be the wrong course of action.

- Social factors: there may be cultural values to consider.

- Religious views: the person's religious beliefs may be important in the choice.

- Financial reasons: the local health service may not have the financial resources for some treatments. Should the patient consider private treatment?

Sometimes the choice is straightforward. For example, ibuprofen can upset some people's stomach, so they will choose another painkiller such as paracetamol.

At other times the choice is difficult: whether to agree to a risky surgical procedure or not.

Some factors are out of the control of the individual and the choice of treatment is down to the medical professionals. This is especially the case where government or personal funding is an issue.

Know more

The ideal treatment for a patient may be very expensive. It may not be available, because of limited NHS resources.

 ...religion and medicine

Depression can be treated by a number of different therapies or a combination of therapies

Treatment case study

P5

Depression can be a debilitating condition. There are several different ways of treating depression. Some are drugs-based therapies (**antidepressants**) and some are based on **cognitive therapy** or other forms of **psychotherapy**. There is no right or wrong way; the decision needs to be made by consultation with a medical professional. The severity of the symptoms needs to be assessed, the type of depression diagnosed and then individual circumstances and choice of treatment need to be discussed.

A combination of drugs and 'talking therapies' may be chosen. The treatment plan should take into account what therapies are available locally and how long the waiting list is. The waiting time for psychological therapies is often very long. Availability is limited because the course of treatment can be lengthy, and so costly for the NHS. The patient may decide to pay for private treatment to speed things up.

The patient may not want to take medication; another patient may not favour psychotherapy.

If antidepressants are prescribed there could be side-effects; if they are not prescribed the person's condition could get worse. The effect of the person's depression on their family and on their work needs to be considered. The patient may even become suicidal.

The factors in the choice of treatment are clearly medical, financial, individual and social.

Research

Research other treatments (for example abortion or IVF treatment) where personal reasons might influence the decision to accept the treatment or not.

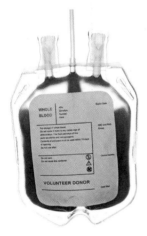

This bag of blood could save a life but might be refused

Reasons not to have a treatment

D4

The reasons for someone to refuse a treatment might be very personal and important. A blood transfusion, for example, is a type of replacement therapy. Either whole blood or specific components of blood is given to the patient. There are some religions that do not agree with this type of therapy. This means that even if the blood transfusion would be life-saving, the treatment should not be administered without the consent of the patient or their family. This has led to controversy and debate, particularly regarding life-saving treatment for children from families holding these beliefs.

Think about

Why might a person with a terminal illness (one they will not recover from) refuse some forms of treatment?

Discuss

Some people have to make extremely difficult decisions, for example whether to continue life support treatment for a loved one, if there is no chance of recovery. Discuss the ethical pros and cons in such a situation.

...ethics and medicine

The cost of treatment

Your assessment criteria:

P5 Identify other factors affecting the choice and availability of treatments to patients

M5 Describe controversial decisions in prescribing treatments

D5 Explain why decisions to give prescription drugs to some and not to others are always controversial

The postcode lottery

The 'postcode lottery' is a phrase adopted by the media to describe some controversial decisions about treatments for some serious conditions. Depending on what area you live in, someone's treatment may or may not be funded by the NHS. These stories are often big news as they can affect someone's chances of survival.

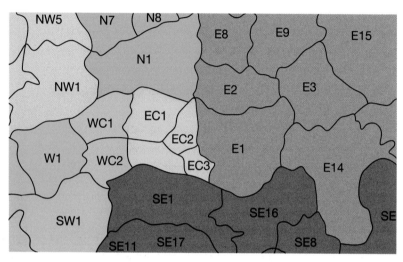

Your postcode could determine the treatment you get

Know more

Some anti-cancer drugs cost up to £4000 per month.

P5 | What are the factors involved?

Some patients do not have the same access to vital drugs and treatments as others in a different part of the country. The key factor determining treatment in this instance is financial.

All **NHS Primary Care Trusts (PCTs)** have allocated **budgets**. This means they have a set amount of money to work with. How they spend the money depends on their choice of priorities. This varies considerably from one part of the country to another.

For example, figures quoted in the *Daily Telegraph* in 2008 alleged that in one part of the country a PCT spent £118 per head on cancer treatments, but a different PCT spent just £47 per head.

Know more

Some of the treatments that could be prescribed might not save a person's life but might prolong it or make the person feel better.

 ...postcode lottery healthcare

Discuss

Is it ethical to deny treatment for financial reasons?

Controversial decisions

M5

Cancer is a condition that can be life-threatening. It can affect any part of the body and is treated with a combination of therapies, including:

- surgery: removal of cancerous growths called **tumours**
- chemotherapy: giving drugs to kill cancerous cells or stop them spreading
- radiotherapy: use of radiation to reduce tumours or kill cancerous cells.

New treatments are being discovered all the time that could offer the chance of curing a cancer or prolonging a life, but they are expensive.

If a PCT has a fixed budget, it will need to decide on financial grounds as to whether they can offer the new treatment. If a new drug prolongs the life of a cancer patient by a few months, can it be considered **cost-effective**?

For example, Taxotere (also known as Docetaxel), is a chemotherapy drug proven to reduce pain and extend the lives of those in the advanced stages of prostate cancer. The problem is that each treatment costs around £1000 and patients need up to 10 treatments. So, although prostate cancer kills one man every hour in the UK, hard and often controversial decisions need to be made.

Sometimes very expensive drugs are not licensed to be used by the NHS. This means that NHS doctors are not 'allowed' to prescribe them.

Research

Find out about other treatments that are, or have been, controversial because of their cost.

Should expensive new treatments be made available to everyone?

What are the implications?

D5

Some therapies do not cure the condition but make it more manageable by reducing pain and other symptoms. Some therapies can prolong the life of a patient. From the point of view of the patient and his or her family and friends, the therapies are considered vital. It is extremely distressing if the treatment is denied.

From the perspective of the organisation offering the therapy, it costs money. Where money is in short supply, it may not make sense to offer an expensive treatment if it does not cure the disease.

- On the one hand, the considerations are clinical and emotional.
- On the other hand, the considerations are clinical and financial.

So any decision made about therapies offered is going to be a controversial one.

 ...docetaxel chemotherapy

To achieve a pass grade, my portfolio of evidence must show that I can:

Assessment Criteria	Description	✓
P1	Carry out investigations into biological and physical procedures used to diagnose illness	
P2	Carry out investigations into the scientific principles of therapeutic drugs used to treat given illnesses	
P3	Carry out investigations into the scientific principles of physical therapies used to treat given conditions	
P4	Identify the general risks of specified treatments	
P5	Identify other factors affecting the choice and availability of treatments to patients.	

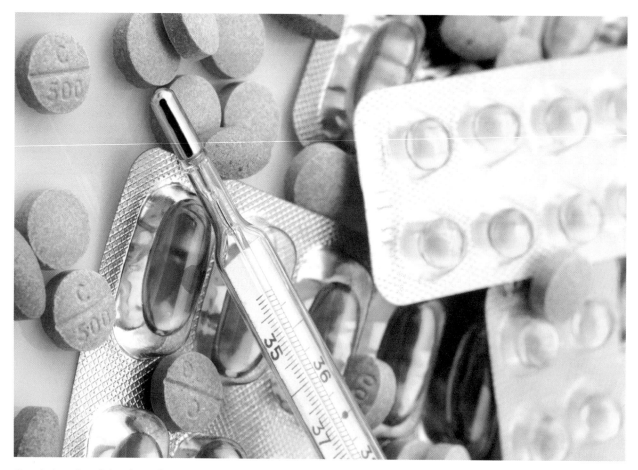

Our choice of medicine depends on a number of factors; it's important to choose wisely

To achieve a merit grade, my portfolio of evidence must show that I can:

Assessment Criteria	Description	✓
M1	Explain the scientific principles underlying the biological and physical procedures used to diagnose illness	
M2	Describe how the therapeutic drugs would be used to treat given illnesses	
M3	Describe how physical therapies would be used to treat given conditions	
M4	Describe, using scientific evidence, the particular risks involved in specified types of treatments	
M5	Describe controversial decisions in prescribing treatments.	

To achieve a distinction grade, my portfolio of evidence must show that I can:

Assessment Criteria	Description	✓
D1	Evaluate the advantages and disadvantages of using biological and physical procedures to diagnose illness	
D2	Explain why the actions of therapeutic drugs are used to treat given illnesses	
D3	Assess, using scientific and other evidence, which physical therapies are effective in the treatment of conditions	
D4	Explain why some individuals may choose not to take advantage of all types of available treatments	
D5	Explain why decisions to give prescription drugs to some and not to others are always controversial.	

Unit 17 Chemical analysis and detection

Know the reagents and techniques used to analyse a variety of chemical compounds

- Chemical analysis of water supplies is used to determine whether they contain poisonous heavy metals, excessive levels of salts or other chemicals such as pesticides

- Scientists analyse unknown chemical substances using simple chemical reagents and sophisticated instruments

- When analysing chemical substances, it's important to identify hazards and assess risks; it may not be possible to eliminate all risk, but they can be limited

Be able to classify compounds according to their pH

- Scientists need to test how acidic or alkaline a substance is, ranging from the air, water and soil around us, to products we use such as cosmetics

- Some acids can dissolve metals; a mixture of concentrated nitric and hydrochloric acids was called aqua regia, or royal water, because it could dissolve even gold

- Alkalis have been used for at least five thousand years to make soap

Be able to show how chromatography is used to analyse materials

- The term chromatography means 'colour writing', coming from the Greek word chroma, meaning colour

- The first proper use of chromatography was by a Russian botanist who used it to separate mixtures of plant pigments

- Gas chromatography using electron capture detection is so sensitive that it's used to detect compounds such as insecticides at the picogram level – that's 10-12 g

Be able to detect different chemicals in unknown compounds

- Is the white powder that has been confiscated a harmless household substance or to be used as a poison?

- Has part of a legal document been written using a different ink, and therefore forged?

- What ingredients does your health drink contain?

- In this LO, you will be analysing substances to find out

Inorganic chemicals

Your assessment criteria:

P1 Identify the reagents needed to analyse inorganic chemicals

M1 Describe the hazards associated with the reagents needed to analyse inorganic chemicals

D1 Explain how to avoid the risks associated with analysing inorganic chemicals

Cornish Sea Salt

The Cornish Sea Salt Company produces its table salt straight from the sea. Water is removed from the sea and filtered. The water is then evaporated, using energy-efficient technology, to leave crystals of pure, white salt.

This salt is extracted direct from the sea

P1

What is salt?

The main chemical compound we get when we evaporate sea water is sodium chloride. It is an **inorganic compound**. It has the formula NaCl. Inorganic compounds are chemical compounds that do not contain carbon.

Sodium chloride is formed by joining an atom of the element sodium to an atom of the element chlorine. When the atoms combine, sodium, a metal, gives one of its electrons to chlorine, a non-metal. This is an ionic bond.

An atom contains an equal number of electrons and protons. After forming an ionic bond, the sodium has lost an electron. So the bonding process leaves the sodium atom with an overall positive charge. The chlorine atom, having gained an electron, becomes negatively charged. We call these charged particles ions. The sodium ion is represented Na^+ and the chloride ion is Cl^-.

Ions that are positively charged, for instance metal ions and hydrogen ions, are called cations. Negatively charged, non-metal, ions are called anions. The table shows the names of some common chemicals and their cations and anions.

Name	Formula	Cation	Anion
sodium chloride	NaCl	Na^+	Cl^-
magnesium oxide	MgO	Mg^{2+}	O^{2-}
copper sulfate	$CuSO_4$	Cu^{2+}	SO_4^{2-}
sodium hydroxide	NaOH	Na^+	OH^-

Know more

Organic compounds contain carbon and come from natural sources (see Unit 4). Inorganic compounds are generally defined as chemicals that do not contain carbon, but carbon dioxide and carbonates are also classified as inorganic.

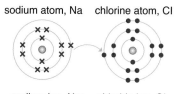

sodium atom, Na   chlorine atom, Cl

sodium ion, Na^+ chloride ion, Cl^-

Sodium and chlorine bond ionically

...ions in solution

Silver nitrate is hazardous

Warning
Harmful chemicals

Evaporating seawater

Seawater contains many different chemical ions in solution. When we evaporate seawater, a series of inorganic compounds crystallise out. The chemicals that crystallise out depend on the concentration of the different ions and their solubility in water.

Testing for chloride ions

Chloride ions can be detected, and their concentration measured, using a chemical reagent called silver nitrate. A chloride is an example of a **halide** – a salt produced from a group 7 element, in this instance, chlorine. Silver nitrate can also be used to detect other halides in seawater, including bromides and iodides. When you add silver nitrate to a halide, a solid called a **precipitate** is produced. The colour of the precipitate, and its solubility when ammonia solution is added, depend on which halide is present.

Halide	Reaction	Notes
bromide	A cream precipitate of silver bromide is obtained.	The precipitate is soluble in *concentrated* ammonia solution.
chloride	A white precipitate of silver chloride is obtained.	The precipitate is soluble in *dilute* ammonia solution.
iodide	A yellow precipitate of silver iodide is obtained.	The precipitate is *insoluble* in *concentrated* ammonia solution.

Research

Find out what the hazards are when using silver nitrate solution.

Think about

Hazards can be biological, chemical or physical. Think about the hazards in your school laboratory. Make a list of five of them.

Assessing hazards

Silver nitrate is a chemical hazard. A hazard is something that could cause you harm. When you use chemicals in your practical analyses, it is important that you know what the hazards are.

Reducing risk

A **risk** is the possibility of harm being done to you by a hazard. It is not always possible to get rid of all the hazards when carrying out an experiment, but it is possible to reduce the risks. To reduce a risk, you need to do a risk assessment.

- First you identify the hazard and the associated risks.

- Then you look at ways of minimising these risks.

When you use chemicals in your practical analyses, you need to know how to reduce the risks.

Research

Find out how to reduce the risks when using silver nitrate solution. What precautions must you take?

...risk assessment in school science

Analysing cations

Is this safe to drink?

You can be fairly certain that tap water in the UK is safe to drink. But what about bottled water?

'Mineral water' is often extracted directly from natural springs, and contains lots of dissolved chemicals. Scientists working for the bottling company need to test the water to find out just what is in it.

P2 Analysing mineral water

Look at the label here from a bottle of mineral water. It tells you the different chemicals that are present in the water and how much of each chemical there is in each litre.

Which metal ions are present in the greatest quantity? What is the concentration of potassium ions in the mineral water?

Flame tests

There are a few simple tests for identifying chemical ions. Mineral water is likely to contain salts of the metals calcium, potassium and sodium. To detect these metal cations we need to evaporate the mineral water and then carry out a flame test on the solid that's left.

1 Put a wire loop in a Bunsen flame to clean it.
2 Dip the loop in hydrochloric acid.
3 Use the wire loop to pick up a tiny crystal of the salt.
4 Hold the loop in a pale blue Bunsen flame and observe the colour change of the flame. This colour can tell you the type of metal ion that's present in the sample.

Any trace of sodium produces a bright yellow flame

Metal ion present	Colour of flame
calcium	red–orange
copper	green–blue
potassium	lilac
sodium	yellow

Carbonated Natural Mineral Water	
Typical analysis	mg per litre
Calcium	25.6
Magnesium	6.4
Potassium	<1.0
Sodium	6.4
Bicarbonate	98.3
Sulphate	10.1
Nitrate	<2.5
Fluoride	<0.1
Chloride	6.8
Silicate	7.6
Dry rediue at 180 °C	109.1
pH	4.6

Know more

Most of the chemicals found in bottled water are inorganic chemicals.

Know more

Dipping the loop in hydrochloric acid first means that a chloride of the metal ion in the sample is formed. Metal chlorides are easier to vaporise than other metal salts, and so work better in flame tests.

...cations and anions table

Some cations can be identified using the sodium hydroxide test

Research

*Chemists at the mineral water company use **chemical instrumental tests** when analysing the water. These are based on 'emission spectroscopy'. Find out how they improve the analyses.*

The sodium hydroxide test

P1

Another test for metal cations is the **sodium hydroxide test.** Just add a few drops of sodium hydroxide solution to the water and watch what happens. If a precipitate forms, its colour will tell you what type of salt or mineral is dissolved in the water.

Metal ion present	What happens when sodium hydroxide is added
calcium	white solid precipitate forms
copper	blue solid precipitate forms
iron(III)	brown solid precipitate forms
lead	white solid precipitate forms that goes when more sodium hydroxide is added
potassium	no change
sodium	no change

Explaining the sodium hydroxide test

M2

We can identify some of the cations present in our sample by the chemical reaction between the sodium hydroxide and the metal ion. A precipitate forms when ions in solution combine to form an **insoluble salt**. We can show this using an **ionic equation**.

$$Cu^{2+} \quad + \quad 2OH^- \quad \rightarrow \quad Cu(OH)_2$$

| copper ions (*in sample solution*) | + | hydroxide ions (*in sodium hydroxide solution*) | → | copper hydroxide (*solid*) |

How reliable are these tests?

D2

If we obtain a white precipitate with the sodium hydroxide test, we can't be absolutely sure of the cation that's present. Many metal ions give a white precipitate with sodium hydroxide. Some metal ions don't react at all, so they may be present too.

Likewise, not all metal ions give a change of flame colour. And any traces of sodium in a sample will turn a flame yellow and hide colours produced by other metal ions.

Cobalt blue glass filters out the yellow colour from any sodium ions. This makes it easy to detect other ions such as potassium.

Testing for anions

Microbrewing

Steve is a microbrewer at Buntingford Brewery. He uses water from a natural well at the brewery site when he brews his beer. It's important for Steve to know exactly which ions are present in the water he uses.

P1

Testing for anions

Steve knows that certain anions – sulphate and chloride – will give the beer a good flavour. He knows that others, such as carbonate and hydrogencarbonate, will hinder the production of beer.

Testing for chlorides

The test for chlorides is very simple. You just add a few drops of silver nitrate solution, acidified with nitric acid, to the sample to be tested. A white solid precipitate forms if chlorides are present. This precipitate darkens as it's exposed to light.

Testing for sulfate

To test for sulfate in a sample, just add a few drops of barium chloride solution, acidified with hydrochloric acid. A white solid precipitate this time shows that a sulphate is present.

Testing for carbonates and hydrogencarbonates

When an acid is added to a sample containing carbonate or hydrogencarbonate ions, the sample fizzes as it gives off carbon dioxide gas. We just need to check that the gas given off is carbon dioxide.

Adding hydrogen chloride to limestone (which contains carbonate)

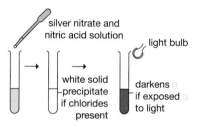

silver nitrate and nitric acid solution

light bulb

white solid precipitate if chlorides present

darkens if exposed to light

Testing for chloride ions

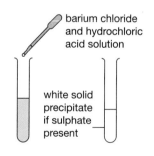

barium chloride and hydrochloric acid solution

white solid precipitate if sulphate present

Testing for sulfate ions

Know more

High concentrations of the cations calcium and magnesium in water are also good for brewing. With chloride and sulfate ions, they make the water acidic. Acidic water creates ideal conditions for yeast to grow and for producing beer.

...cations and anions table

Know more

Another important anion is the hydroxide ion, OH^-. Hydroxide ions are formed when certain soluble bases (alkalis) dissolve in water. They are best tested for using an indicator, such as universal indicator. You will learn more about testing for alkalis later in this unit.

Testing for nitrates

All the wastes from Buntingford Brewery are purified naturally using plants called reeds in a reedbed system. Every so often, Steve checks nitrate levels in the reedbed system to make sure that the brewery wastes have been successfully treated.

Brewery waste water is treated by discharging it into reedbeds

There are several chemical tests for nitrates. The easiest way is to use a special nitrate-testing kit. These kits contain either plastic strips or reagents that change colour if nitrates are present. Any colour change is compared with a colour chart. The colour change will indicate the concentration of nitrates present in the water.

Think about

Write the ionic equation for the reaction between barium chloride and sulfate ions. Barium ions are written Ba^{2+} and sulfate ions are written SO_4^{2-}.

The chemical reactions involved

Precipitates form when ions in solution combine to form an insoluble salt. We can show the reaction between silver nitrate and chloride ions by writing an ionic equation:

$$Ag^+ \quad + \quad Cl^- \quad \rightarrow \quad AgCl$$

silver ions	chloride ions	silver chloride
(*in silver nitrate* + solution)	(*in sample* solution)	$\rightarrow$ (*solid*)

How reliable are these tests?

These tests are quick and easy to carry out, but can we be sure of our identifications of the anions present? Could several different ions give similar results?

Most scientists would not rely on these tests in their labs. They must be sure of their identifications. And there are many anions for which there are no chemical tests available.

Steve sends water samples to an **analytical chemist**, who uses **ion chromatography** to analyse them. This technique is not only reliable but is quantitative. It tells Steve the concentrations of the different ions in the water.

Think about

Find out how ion chromatography works.

Analysing gases

What's that gas?

Lily has been testing for anions in the school laboratory. There is a gas produced during one of the tests. She needs to test the gas to see if it is carbon dioxide.

P2

Testing for gases in the school lab

The gases that you are going to test for are carbon dioxide, hydrogen and oxygen. These are all invisible gases. This makes it very difficult to know if the test tube is full of the gas or not. You will need to carry out the tests quite quickly to ensure that the gas has not escaped from a test tube.

Testing for carbon dioxide

Carbon dioxide is given off when acid is added to a sample containing carbonate or hydrogencarbonate anions. When carbon dioxide gas is bubbled through lime water, the **lime water** changes from clear to milky-coloured.

Testing for hydrogen

Hydrogen gas is less dense than air so will quickly float out of an upright test tube. If the test tube contains only a small sample of gas, you can ensure that the gas stays in the test tube by holding the test tube upside down.

Hydrogen is a gas that burns. When mixed with oxygen in the air it forms an explosive mixture. You can test for hydrogen gas by introducing a burning wooden spill into a test tube of hydrogen. The hydrogen will explode with a loud squeaky pop.

Exploding a test tube of hydrogen is quite safe, but **do not try it** with larger quantities of hydrogen.

Testing for oxygen

Oxygen is a gas that will relight a glowing wooden spill. To test for oxygen, light a wooden splint and then blow it out. The wood should be still glowing. Introduce the end of it into a test tube of the gas. If the unknown gas is oxygen, the glowing splint will relight and burst into flame.

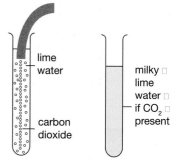

lime water

milky
lime
water
if CO_2
present

carbon dioxide

Testing for carbon dioxide

Know more

You can try this by breathing out through a straw into a beaker of lime water. The carbon dioxide gas in your breath gradually turns the lime water milky.

Testing for oxygen

...chemistry gases test

Limestone is calcium carbonate. It is dissolved by acid rain

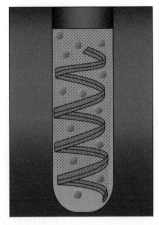

Magnesium metal dissolves in acid. Hydrogen gas is produced

M2

How do we know which gas to test for?

We can make a reasonable assumption of the gas to be tested for if we know about the chemistry of the chemical reaction that's occurring.

- All carbonates react with acids to produce carbon dioxide.

$$CaCO_3 + H_2SO_4 \rightarrow CaSO_4 + CO_2 + H_2O$$

calcium carbonate + sulfuric acid → calcium sulfate + carbon dioxide + water

- Carbon dioxide is also produced as fuels burn.

$$CH_4 + 2O_2 \rightarrow CO_2 + 2H_2O$$

methane + oxygen → carbon dioxide + water

- Hydrogen is produced when some metals react with acids.

$$Mg + H_2SO_4 \rightarrow MgSO_4 + H_2$$

magnesium + sulfuric acid → magnesium sulfate + hydrogen

- Very reactive metals can react with water to produce hydrogen.

$$2K + 2HOH \rightarrow 2KOH + H_2$$

potassium + water → calcium hydroxide + hydrogen

- Oxygen is produced when chemicals called **oxidising agents** break down.

$$2H_2O_2 \rightarrow 2H_2O + O_2$$

hydrogen peroxide → water + oxygen

Think about

Not all metals react with acids to produce hydrogen. Explain why some do not.

D2

How reliable are the tests?

There may not be enough gas to give a positive result using these tests. There may be too much gas – you couldn't test for hydrogen in a large volume of gas using a lighted splint.

Very often, scientists need to know the quantity or concentration of a gas, not just whether or not the gas is present. Scientists therefore use sensors to test for gases. Sensors are important in industrial reactions involving gases, and in environmental monitoring.

Hydrogen sensors will also become important if the number of hydrogen-powered cars increases.

Acids, bases and alkalis

Useful chemicals

Acids, bases and alkalis are all around us. The soil and the water where you live may be acidic or alkaline. We use acids, bases and alkalis in our homes.

P3

What are acids, bases and alkalis?

Acids were first identified as chemicals that were sour. Many weak acids give different foods the sour flavours we're familiar with. But strong acids, or weak acids in high concentrations, are so **corrosive** that it is dangerous to try to taste them.

Later on, scientists recognised that all acids contained one or more hydrogen atoms.

Formula of acid	Name	Example
H_2CO_3	carbonic acid	fizzy lemonade
CH_3COOH	ethanoic acid	vinegar
HCN	hydrocyanic (prussic) acid	poison or Zyklon B
HCl	hydrochloric acid	stomach
$HOOCCH_2CHOHCOOH$	malic acid	apples
HCOOH	methanoic acid	ants
HNO_3	nitric acid	fuming conc. nitric acid
H_2SO_4	sulfuric acid	chemical industry

Alkalis were probably first recognised as the corrosive substances obtained from the ashes of fires. In combination with tallow (animal fat), they were used to make soap. Solid alkalis are called **bases**.

The acids that these foods contain give them a sour taste

Know more

Liquids are acidic, alkaline or neutral. Bases are solids and include the oxides, hydroxides and carbonates of metals.

🔍 ...acids and bases

Know more

Litmus is extracted from a lichen.

Acid–base indicators

It is not always easy to tell whether liquids are acids or alkalis. But nature has given us a way. Some flowers, such as hydrangeas, are different colours depending on the soil that they are growing in. If the soil is slightly acidic, hydrangea flowers are blue. If the soil is slightly alkaline, the flowers are pink. If the soil is neutral, the flowers are creamy coloured.

Hydrangea flowers are blue in acidic soil, and pink in alkaline soil

Scientists have discovered dyes that can be extracted from plants that change colour in acids and alkalis. One of these is called **litmus**. Unlike hydrangea petals, litmus turns:

* red in acid

* blue in alkali.

Chemicals, such as litmus, that change colour depending whether a liquid is an acid or an alkali, are called **indicators**.

The pH scale

It is important that we know just how strong an acid or an alkali is. If it is only weakly acidic or alkaline, it should be safe if we get any our skin. But strong acids and strong alkalis can damage our skin very quickly, giving us serious burns.

Scientists need a measure of acidity that's easy to use. The **pH scale** goes from 0 to 14 and tells us just how strong an acid or alkali is.

* pH 1 is a very strong acid.

* pH 14 is a very strong alkali.

* pH 7 is **neutral** (neither acid nor alkali).

The further the number goes away from pH 7, the stronger the acid or the alkali gets. Vinegar has a pH of about 3 or 4. This makes it a weak acid. Sodium hydroxide solution has a pH of around 13. This makes it a strong alkali.

0	
1	— Battery acid
2	— Lemon juice
3	— Vinegar
4	
5	
6	
7	— Milk
8	
9	— Baking soda, Seawater
10	
11	— Milk of Magnesia
12	— Ammonia
13	— Sodium hydroxide solution
14	

Increasing Acidity ↑

Neutral

Increasing Alkalinity ↓

Know more

Pure water is pH 7 so it is not an acid or an alkali. It is neutral.

🔍 ...pH scale

Testing for acids and alkalis

Your assessment criteria:

P3 Carry out experiments to classify compounds according to their pH

Practical

D3 Explain, with examples, the difference between an acid, a base and an alkali

pH in the environment

Justin is an environmental officer. He measures the pH of water and soil samples. It's important that he can detect any small change in pH over a period of time. This would affect plant and animal species.

P3 Measuring pH

Litmus solution or litmus paper will tell us whether something is acidic or alkaline. But it's not of much use to Justin, who needs to know the pH number of his samples. He could choose an indicator that changes to many different colours across the pH range. This indicator is called **universal indicator**.

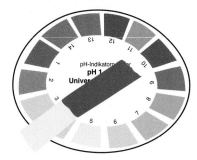

The colours obtained with universal indicator

You can use universal indicator paper or solution to test many different household substances, such as tea, coffee, milk, water, bleach, soap and washing-up liquid. It can also be used in the laboratory to test chemicals such as sodium hydroxide and hydrochloric acid.

Universal indicator is a quick way of measuring the approximate pH of a liquid, but it won't give an exact pH. It's also difficult to use if the solution being tested is coloured. And for a scientist like Justin, doing environmental testing in the field, it's not always convenient to carry bottles of indicator solution around.

Know more

Universal indicator is a mixture of several different indicators. It contains methyl red, bromothymol blue, thymol blue and phenolphthalein. These indicator chemicals change colour across different pH ranges.

...acid–base indicator

P3

Using a pH meter

Digital **pH meters** are a much more accurate way of determining the pH of a solution. They give an exact numerical readout of the pH. They are easy to use, but they are expensive to buy and can be quite fragile.

pH meters have to be calibrated before use. To calibrate a pH meter, readings are set on the meter using liquids whose pH is known. A liquid of known pH is called a **buffer**. Calibrating the pH meter means that the readings we'll get from it will be accurate. In other words, they will be very, very close to their true pH value.

The meter consists of a probe that is placed into the solution, and a digital readout that displays the pH. There are various types of meter and they all work in a slightly different way. You will need to check with your teacher how to use and calibrate the device you are using.

A digital pH meter

pH	Concentration of hydrogen ions /mol dm^{-3}	
0	10^0	1
1	10^{-1}	0.1
2	10^{-2}	0.01
3	10^{-3}	0.001
4	10^{-4}	0.0001
5	10^{-5}	0.00001
6	10^{-6}	0.000001
7	10^{-7}	0.0000001
8	10^{-8}	0.00000001
9	10^{-9}	0.000000001
10	10^{-10}	0.0000000001
11	10^{-11}	0.00000000001
12	10^{-12}	0.000000000001
13	10^{-13}	0.0000000000001
14	10^{-14}	0.00000000000001

Think about

Explain the difference between an acid and an alkali.

D3

What does pH measure?

The pH scale tells us the concentration of hydrogen ions in a liquid. Hydrogen ions are hydrogen atoms that have lost an electron.

- At pH 1, the concentration of hydrogen ions is 10^{-1} mol dm^{-3}, or 0.1 mol dm^{-3}.

- At pH 2, the concentration of hydrogen ions is 10^{-2} mol dm^{-3}, or 0.01 mol dm^{-3}.

So, the concentration of hydrogen ions at pH 2 is one-tenth of the concentration at pH 1, and so on. Look at the table. The pH scale is a logarithmic scale.

You can see that the hydrogen ion concentration in alkalis is extremely low.

...pH meter calibration

Uses of acids, bases and alkalis

Chemistry in the kitchen

Chetna has been investigating acids, bases and alkalis in her science course at school. She buys a pH-testing kit from a garden centre and decides to test some of the chemicals in the kitchen cupboards.

Your assessment criteria:

P3 Carry out experiments to classify compounds according to their pH

Practical

M3 Explain the uses of the classified compounds in the laboratory and home

D3 Explain, with examples, the difference between an acid, a base and an alkali

P3 Testing different chemical compounds

At school, Chetna has tested the pH of a range of different chemical compounds using universal indicator. These have included hydrochloric acid and sulphuric acid, sodium hydroxide and potassium hydroxide.

In the kitchen at home, she tests cleaning products, soap and detergents, some food products and some medicinal products. For the liquids, she pours a small amount into a container, and adds two drops of indicator solution. For solid products, she dissolves the solid in a small amount of distilled water and then tests them in the same way.

Know more

Chetna also tests the pH of the tap water. She finds that it is slightly acidic. This is because carbon dioxide in the air readily dissolves in water. It forms carbonic acid.

M3 What are the uses of acids, bases and alkalis?

The industrial uses of the common acids, bases and alkalis that Chetna has tested in school are shown here.

Sulfuric acid

Fertiliser production, e.g. ammonium sulfate, superphosphate

Cleaning iron and steel by dissolving oxides

Pigment production, e.g. by dissolving pigments from ores

Detergent production

Sodium hydroxide

Soap manufacture

Biodiesel manufacture

Paper manufacture

Cleaning agent, for fats and oil

Chemical analysis

Hydrochloric acid

Manufacturing plastics, e.g. PVC

Pharmaceutical products

Cleaning products

Potassium hydroxide

Soap manufacture

Biodiesel manufacture

Research

Research one of the uses of hydrochloric or sulfuric acid shown here, and one of the uses of potassium or sodium hydroxide. Explain how these uses depend on the way the acid/alkali reacts with other chemicals.

 ...acids and alkalis in the home

Oven cleaner is a strong alkali

M3

Alkaline and basic products in the kitchen

Chetna tests some oven cleaner very carefully, wearing gloves. It is a strong alkali. When she tests soap, and some detergent, these also turn her indicator blue. These alkaline products are designed to 'cut through' and dissolve fat and grease.

The 'antacid' remedies for indigestion in the kitchen cupboard also have a high pH. The active ingredient in antacid tablets is a base. Liquid antacid preparations such as Milk of Magnesia are alkalis.

Acidic products in the kitchen

When Chetna tests the 'eco-friendly' disinfectant in the kitchen, she finds it to be acidic. And so is the product her father uses for removing **limescale** from the kettle.

Limescale remover is acidic

Chetna finds some vitamin C tablets in the cupboard. She grinds them up and adds water. They are acidic. Many of the food products she tests are neutral. But pickled foods are acidic.

Think about

Shampoo is a detergent. Some shampoos are buffered to pH 5.5. Why do shampoo manufacturers do this?

Research

The chemical reaction by which soap is produced is similar to the reaction of sodium hydroxide (oven cleaner) with fat. Find out more about the manufacture of soap.

Think about

Limescale is mostly calcium carbonate. This is a basic compound that precipitates out when hard water is boiled. Limescale remover contains phosphoric acid (H_3PO_4). Explain how it works. Write a word equation for the reaction between calcium carbonate and phosphoric acid.

D3

How do the chemical properties of these compounds enable them to work?

The alkali in oven cleaner is sodium hydroxide. Sodium hydroxide and potassium hydroxide are able to react with and dissolve fat.

sodium hydroxide + triolein *(a common type of fat)* → sodium oleate + glycerol

The indigestion tablets in Chetna's cupboard are made from a base called calcium carbonate. Carbonates react with and neutralise excess hydrochloric acid in a person's stomach.

$$CaCO_3 + 2HCl \rightarrow CaCl_2 + CO_2 + H_2O$$
calcium carbonate + hydrochloric acid → calcium chloride + carbon dioxide + water

Milk of Magnesia is an alkaline liquid, magnesium hydroxide, which also neutralises stomach acid.

$$Mg(OH)_2 + 2HCl \rightarrow MgCl_2 + 2H_2O$$
magnesium hydroxide + hydrochloric acid → magnesium chloride + water

🔍 ...uses of acids and bases

Reactions of acids, bases and alkalis

Your assessment criterion:

D3 Explain, with examples, the difference between an acid, a base and an alkali

Making fertiliser

Atiq is making some inorganic fertilisers in the school lab to test the effects they have on some plants he's growing. The two fertilisers he's chosen to make are ammonium sulfate and potassium nitrate.

D3

Neutralisation reactions

The fertilisers that Atiq is making are **soluble salts**. They can be made by reacting acids and alkalis. When an acid and alkali (or base) are reacted together, a salt and water are formed.

$$\text{acid} \ + \ \text{base} \ \rightarrow \ \text{salt} \ + \ \text{water}$$

In the lab, potassium nitrate can be made by reacting the appropriate volume of nitric acid with potassium hydroxide solution. This reaction is a **neutralisation**. The alkali, potassium hydroxide, is neutralised by the nitric acid.

$$\text{nitric acid} \ + \ \begin{array}{c}\text{potassium}\\\text{hydroxide}\end{array} \ \rightarrow \ \begin{array}{c}\text{potassium}\\\text{nitrate}\end{array} \ + \ \text{water}$$

$$HNO_3 \ + \ KOH \ \rightarrow \ KNO_3 \ + \ H_2O$$

Acids are chemicals that form hydrogen ions, H^+, in water. Alkalis are chemicals that form hydroxide ions, OH^-, in water. The ionic equation for a neutralisation reaction is:

$$H^+ \ + \ OH^- \ \rightarrow \ H_2O$$
$$\text{hydrogen ion} \ + \ \text{hydroxide ion} \ \rightarrow \ \text{water}$$

In the reaction between nitric acid and potassium hydroxide, the other ions in solution, K^+ and NO_3^-, will form crystals of potassium nitrate when the water is evaporated off.

So that he knows when the neutralisation reaction is complete, Atiq adds a few drops of an indicator to the flask of potassium hydroxide solution. He uses a burette to add the nitric acid. The burette has a tap so that the acid can be added at a controlled rate. It is also graduated so that Atiq can read off the volume of acid that he's used to neutralise the alkali.

Scientists use a technique similar to this to analyse the concentrations of chemicals in solution. The technique is called **titration**.

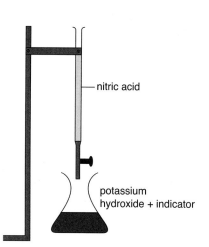

— nitric acid

potassium hydroxide + indicator

Apparatus used for the neutralisation reaction

...titration indicators

D3

Think about

Sodium ethanoate is used in some brands of salt and vinegar crisps to give them their flavour. It is a salt of a strong base and a weak acid. Will its pH be below 7, neutral or above 7?

Research

Find out how titration is used to analyse the concentrations of chemicals in solution. Which types of scientist would need to do this?

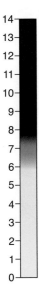

The colour of bromothymol blue at different pHs

Research

Which indicator would be best for a weak acid–strong base titration?

Strong and weak acids and bases

Potassium hydroxide is a strong base, and nitric acid is a strong acid. When the potassium hydroxide solution is neutralised, the pH of the liquid will be pH 7.

But not all acids and bases are neutralised at pH 7. Atiq also makes some ammonium sulfate, by reacting ammonium hydroxide and sulfuric acid. Ammonium hydroxide is a *weak* base, and sulfuric acid a *strong* acid. The salt that's made, ammonium sulfate, is acidic. Although the ammonium hydroxide is neutralised by the reaction, the final pH of the ammonium sulfate solution is 5.5.

Using the right indicator

Atiq has a choice of indicators to use when making his fertilisers. When making ammonium sulfate, he chooses one called methyl red.

The colour of methyl red at different pHs

Methyl red indicator is a yellow colour in alkaline solution, and red in acid. At the point when the alkali has been neutralised, at the pH of ammonium sulfate (5.5), the colour of the indicator is orange.

For the strong acid–strong base reaction that makes potassium nitrate, Atiq decides to use bromothymol blue. At pH 7 this indicator is green.

Other indicators are also suitable for strong acid–strong base titrations. This is because the pH change is very sharp when a very small volume of acid or alkali is added, as the graph here shows.

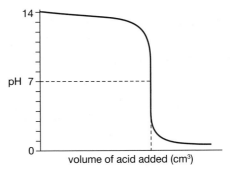

The change in pH for a strong acid–strong base reaction

Chromatography

Colours in black ink

When a page of writing gets wet, sometimes different colours begin to separate out. Even though the ink is black, it may be made from several different colours that have been mixed together.

P4

Paper chromatography

Chromatography is a technique used to separate substances from mixtures, using their colours. Try the following.

1 Take a piece of filter paper and place a small drop of black ink in the centre of the paper. Allow the drop to dry.

2 Cut a strip about 5 mm wide, from the outside of the filter paper towards the centre, stopping just short of the dried ink drop.

3 Bend the paper strip at right angles to the filter paper. Place the filter paper over a beaker containing about a 3 cm depth of water, so that the end of the paper strip is below the level of the water.

The water soaks up the paper strip and spreads out through the filter paper, carrying the ink with it. Because different colour dyes travel at different speeds, the colours in the black ink begin to separate out. This shows all the different dyes that were used to make the black ink.

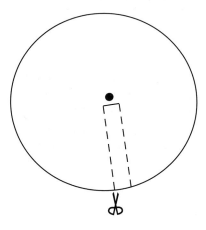

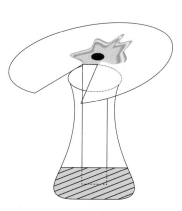

Paper chromatography

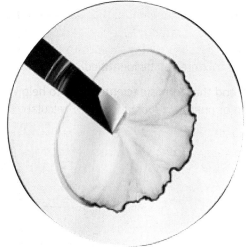

The finished chromatogram should look something like this – the dyes in the ink have separated

Know more

Not all the chemical substances that you'll separate by chromatography are coloured. So you will need a chemical called a visualisation agent to show them up.

🔍 ...paper chromatography technique

Advantages and disadvantages of paper chromatography

D4

Paper chromatography is a simple and fairly quick procedure. But the chromatography paper, when it's wet, can be quite difficult to handle. And the chemical nature of the paper itself is not perfect for separating mixtures of certain types of chemical compound.

The major disadvantage of paper chromatography (and in fact all forms of chromatography) is that, unlike the chemical tests you have carried out, it does not directly identify the substances. Nor does it tell you the proportion of each substance in the mixture.

Identifying the chemicals separated by chromatography

Sometimes it's not necessary to identify the compounds you've separated. A comparison may be sufficient.

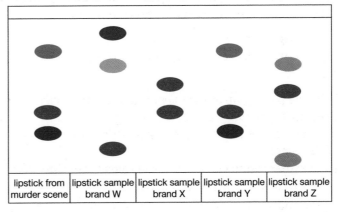

lipstick from murder scene	lipstick sample brand W	lipstick sample brand X	lipstick sample brand Y	lipstick sample brand Z

A forensic scientist has identified the brand of lipstick found at a murder scene without identifying individual dyes within it

If identification is needed, scientists can also use *standard reference solutions* (or *standards*). These are run alongside the sample being analysed. By comparison with the standards, the unknowns in a mixture can be identified.

Another method that forensic scientists use to help with the identification of unknown substances is to calculate **Rf values.**

When you made your chromatogram, you will have noticed that different colours move at different rates compared to the water. If you divide the distance that each colour moves by the distance that the water moves (the **solvent front**), you can calculate the Rf (**retention factor**) value for each substance. Different colours will have different Rf values.

A forensic scientist can compare the Rf values obtained with values printed in scientific papers and textbooks. See the example on the left.

Research

Sometimes a better separation of a mixture can be obtained by two-dimensional (2-D) chromatography. How does 2-D chromatography work?

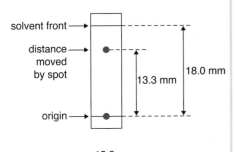

$$Rf = \frac{13.3}{18.0} = 0.74$$

Calculating the Rf value

How chromatography works

Your assessment criteria:

M4 Demonstrate how chromatography works to separate materials

D4 Evaluate the advantages and disadvantages of using chromatography to analyse materials

How do mixtures separate?

To use and analyse chromatography effectively, it helps to know how it works. It is all to do with the attraction between the chemicals in the mixture, in the paper and in the water.

M4

The stationary and mobile phases

When you made your chromatogram using filter paper, the water, or **mobile phase**, acted as a solvent. The water moved through the paper, or **stationary phase**, carrying the different colours with it. The different dyes in the ink moved at different speeds, so they began to separate.

How quickly each dye substance moves through the stationary phase depends on how strongly the molecules of the substance are attracted to the stationary phase and to the mobile phase.

- The stronger the attraction that a dye has to the stationary phase (paper), and the weaker the attraction to the mobile phase (water), the slower it moves.

- For dyes more strongly attracted to the mobile phase than to the stationary phase, the quicker they move.

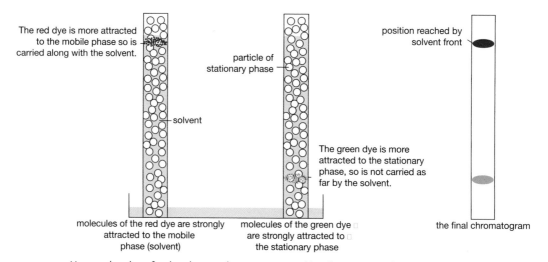

The red dye is more attracted to the mobile phase so is carried along with the solvent.

particle of stationary phase

solvent

position reached by solvent front

The green dye is more attracted to the stationary phase, so is not carried as far by the solvent.

molecules of the red dye are strongly attracted to the mobile phase (solvent)

molecules of the green dye are strongly attracted to the stationary phase

the final chromatogram

How molecules of red and green dye are separated by chromatography.

 ...how does chromatography work?

Using different stationary and mobile phases

D4

When scientists use chromatography to separate mixtures, they rarely use paper as the stationary phase.

Also, they rarely use water as the mobile phase. Simple solvents like water usually don't separate the components of a mixture very effectively. A mixture of solvents, called a **solvent system**, is almost always used. Scientists experiment with different proportions of a range of solvents to find the best solvent system for the job.

For the most effective separation of mixtures, specialist chromatographic techniques are used. The best method for the substances involved is the one that is the most sensitive. This means that minute quantities of chemical compounds in a mixture can be effectively separated.

Thin-layer chromatography (TLC)

In this technique, the stationary phase is a solid coated on a sheet of glass or plastic. This is called a **TLC plate**. Three main types of coating can be used – silica gel, alumina or cellulose. Scientists choose the coating depending on the chemicals they're trying to separate.

Gas chromatography

In gas chromatography, the stationary phase is a liquid coating on a solid, packed into a glass tube. The mobile phase is a gas such as nitrogen or helium.

High-performance liquid chromatography (HPLC)

Here the stationary phase is a solid and the mobile phase a liquid.

Applying a mixture for analysis to a TLC plate

Research

Research some of the applications of gas chromatography and high-performance liquid chromatography.

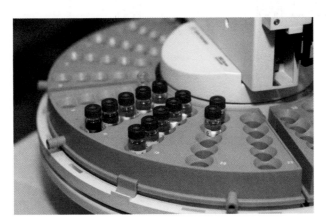

Gas chromatography

Chromatography of plant juices

Your assessment criteria:

P4 Carry out experiments to show how chromatography is used to analyse materials

> Practical

D4 Evaluate the advantages and disadvantages of using chromatography to analyse materials

Wheat grass juice

Leigh is a chemical analyst who works for a chain of juice bars. Her job is to check the nutritional content of the fruit and other plant juices. Today she's checking the wheat grass juice.

P4

The chlorophyll content of wheat grass

'Wheat grass' is the green shoots of wheat seedlings. It contains many nutrients beneficial to human health. The juice is particularly rich in the chemicals that give green plants their colour. These are a range of chemical compounds called **chlorophylls**. Some scientists think that chlorophylls may help to protect people against certain types of cancer.

Leigh grinds up some wheat grass in an organic solvent using a mortar and pestle. She separates the plant pigments using the technique called thin-layer chromatography. The plant pigments separate out on the chromatogram.

She can use chromatography to separate the pigments in a variety of different plants, including seaweeds.

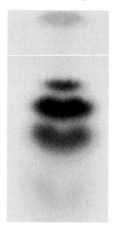

Plant pigments separated by chromatography

Analysing fruit juices

Leigh is also interested in the fruit juices that the company sells. She regularly monitors their amino acid content. She prepares an extract of each fruit juice and uses thin-layer chromatography to separate out the amino acids in each juice. Amino acids are colourless, so Leigh has to spray her chromatogram with a chemical called ninhydrin to reveal them.

Know more

Chlorophylls are not the only type of plant pigments. Others, called accessory pigments, are usually hidden because there's so much chlorophyll in leaves. But in the autumn, when the chlorophyll breaks down, you can see the yellow, red and brown accessory pigments.

Think about

Suggest how Leigh can identify the amino acids in the different fruit juices.

...chromatography amino acids

Making chromatography quantitative

D4

The TLC analysis has told Leigh which plant pigments are in the wheat grass extract, but it hasn't told her how much there is of each. She wants to find out the concentrations of the chlorophylls and other plant pigments in the wheat grass juice. She uses **column chromatography**.

She prepares her wheat grass extract as usual, but this time pours it through a glass column packed with powdered silica gel moistened with solvent.

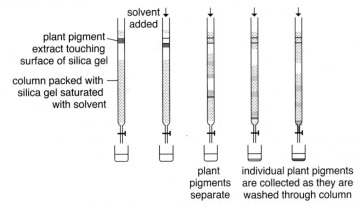

The sample is poured into the column

Leigh then adds more solvent to the column. The solvent washes down the plant pigments. They move at different rates. Leigh collects each pigment as it runs out of the column. By evaporating the solvent, she can find out how much there is of each pigment.

Improving the separation of amino acids

Using TLC, it's very difficult to get a perfect separation of the amino acids in the fruit juices. This is because some of the amino acids have Rf values that are very close to one another. Leigh also needs to find the concentration of amino acids in the different fruit juices. So she uses column chromatography to separate the amino acids. But it's a different type to the one she used for the plant pigments.

This time, she uses high-performance liquid chromatography (HPLC). HPLC enables scientists to identify amino acids in extracts from organisms and calculate their concentrations. The liquid mobile phase is forced through under high pressure. The amino acids are retained in the column for different lengths of time, and can be identified.

Using HPLC, Leigh can be fairly sure of the identity of each amino acid in the sample. If she needs to be absolutely certain, she can connect the chromatography apparatus to an instrument called a mass spectrometer.

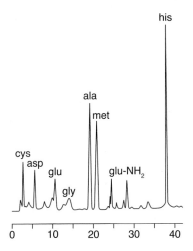

The size of the peaks on the HPLC trace depends on how much of each amino acid is present. A computer analyses the peaks and works out the concentration of each amino acid in the mixture

Analysing unknown compounds

Suspicious substances confiscated

Customs officers at ports and airports sometimes find people carrying suspicious substances. These may be powders, which appear to be household chemicals or food products. They may be what it says on the label, if they are labelled. But they may be something else, or have had some other chemical added. The substances will be sent to a chemical analyst for investigation.

Your assessment criteria:

P5 Carry out experiments to identify chemicals in unknown compounds

> Practical

M5 Explain the scientific principles behind the tests used to identify the chemicals in unknowns compounds

D5 Evaluate the results of the analysis, considering how to improve subsequent experiments

P5

Carrying out chemical analyses

You will be provided with some substances for analysis. These will be unknown chemicals, or mixtures of chemicals. You will use some of the tests that you have been reading about: chemical tests, pH measurements and chromatographic techniques.

Some items that customs officers may consider suspicious

Imagine you are given an unknown white powder. It could be as harmless as chalk or as dangerous as cocaine. First you might carry out the tests for cations: the sodium hydroxide test and the flame test. Then you could carry out the tests for anions; then identify any gases given off. You could also test the pH of the powder. If the powder is a mixture, you could try separating the components using chromatography.

List your results clearly. Some example results are shown on the next page.

🔍 ...chemical analysis of anions and cations

Results of tests on an unknown chemical

	Name of test	Result
cations	sodium hydroxide	white precipitate
	flame test	red–orange
anions	carbonate	fizzes
	chloride	no reaction
	sulfate	no reaction
	nitrate	no reaction
	OH	no reaction
gases	carbon dioxide	lime water turned milky
	hydrogen	no reaction
	oxygen	no reaction

Conclusions

M5

You will need to consider your results and draw conclusions about the identity of the chemicals. You must justify the conclusions you have made. This means that you need to explain the scientific principles behind each test and how this helped you to draw that conclusion. For example:

I know that the substance I analysed is sodium chloride because it gave an orange-yellow flame colour and a precipitate of silver chloride when silver nitrate was added.

Think about

Draw your conclusions from the results in the table above. Determine the name of the unknown chemical.

Evaluating the procedure

D5

You need to be able to evaluate the accuracy of the techniques you have used to analyse the chemicals.

For example, a student got different results from those in the table above when testing the same substance. He found that the flame test gave a yellow flame, not a red–orange flame. This meant he could not decide whether the cation was calcium or sodium. The reason he got this confusing result was that he had either not carefully washed the glassware or not cleaned the flame-testing loop beforehand. One or both had been contaminated with a sodium compound.

You should be able to suggest and explain ways that your technique could be improved. In this case, one way of improving the technique would be to use very clean apparatus.

In a chromatography investigation, one improvement might be to run the chromatogram for longer.

...chromatography analysis

Assessment and Grading criteria

To achieve a pass grade, my portfolio of evidence must show that I can:

Assessment Criteria	Description	✓
P1	Identify the reagents needed to analyse inorganic chemicals	
P2	Describe the techniques needed to analyse inorganic chemicals	
P3	Carry out experiments to classify compounds according to their pH	
P4	Carry out experiments to show how chromatography is used to analyse materials	
P5	Carry out experiments to identify chemicals in unknown compounds.	

By doing chemical analysis you can determine if these products are toxic

To achieve a merit grade, my portfolio of evidence must show that I can:

Assessment Criteria	Description	✓
M1	Describe the hazards associated with the reagents needed to analyse inorganic chemicals	
M2	Explain the results of using these techniques by providing the formula of an unidentified compound	
M3	Explain the uses of the classified compounds in the laboratory and home	
M4	Demonstrate how chromatography works to separate materials	
M5	Explain the scientific principles behind the tests used to identify the chemicals in unknown compounds.	

To achieve a distinction grade, my portfolio of evidence must show that I can:

Assessment Criteria	Description	✓
D1	Explain how to avoid the risks associated with analysing inorganic chemicals	
D2	Evaluate the accuracy of techniques used to analyse inorganic chemicals and explain how they could be improved	
D3	Explain, with examples, the difference between an acid, a base and an alkali	
D4	Evaluate the advantages and disadvantages of using chromatography to analyse materials	
D5	Evaluate the results of the analysis, considering how to improve subsequent experiments.	

Research and investigate

Science affects all our lives. It affects the world around us, by trying to answer big questions such as:

- How did the world get to be the way it is?

- How can we make the world safer for its inhabitants?

- How can we look after it for future generations?

So what do scientists do? They gather **data.** Data is information. Scientists observe, collect information, and use it to find out things about the world.

Reliability

Scientists always need to consider whether the data is ***reliable****. They ask themselves: how confident are we about it? Can we trust it? These are the questions you have to ask yourself when dealing with your data.*

Events and variables

Things change, and this interests scientists. They like to explain changes, so that they can be managed and predicted. They call the things that change **variables**.

Unlinked events

Sometimes things cannot be predicted and managed. This is a **chance occurrence**. While you are brushing your teeth in the morning, a baby is born somewhere in the world – but it is unlikely that these events will be linked.

Linked events

Sometimes, though, events are linked. If they are, the link can be explained. If you exercise for a period of time, your heart will beat faster. Here, one **variable** (how active you are) changes another **variable** (your heart rate). Usually, in science, several variables are involved and linked. If they are linked, they can be explained.

For example...

You might notice that the sandwich you bring from home is drier at lunchtime than it is at breakfast. Why is this? You might decide to investigate whether it has lost moisture. You could then reach a conclusion about how the sandwich changes over time, and why.

Observations

Scientists spend a lot of time making observations. Observations raise questions, and these may lead to an investigation.

Investigations

What is an investigation?

If you change an object, and keep track of how it is different, you are seeing the effect of your actions. You are measuring the amount that it has changed. This is an investigation!

Scientists use a lot of different ways to measure things. If they do it correctly, the results of these measurements can tell them something about the world.

Many investigations have an important real-life purpose, like:

- Searching for new drugs to combat cancer or heart disease

- Trying to make mobile phones that work on the Underground

- Finding alternative energy sources and solutions of global warming

To carry out your investigation you need to think about what your scientific purpose is.

When getting your results, leave room for human error – nobody is perfect.

Ask yourself...

- *What are you trying to find out?*

- *What do you need to do to find this out?*

- *How are you going to do it?*

When you know the answer to these questions, you're ready to start your investigation.

Always be sure what you are trying to find out.

Start investigating

To start, you need to think about the **big question**. What are you trying to find out?

The 'big question' will lead to other questions. Finding a way to answer all these smaller questions will give you a way to answer the big question.

You can answer questions through tests. The tests you use must be **fair tests**. It's usually better to do them in a lab, where it's easier to control variables. (Though that's not always possible, so some tests have to be carried out 'in the field'.)

An investigation must be designed to give data that is reliable and precise – not vague. After your investigation you should think about and write about the reliability of the measurements you have made. You should **evaluate** this reliability in your conclusion.

Display your results

It is useful to display the data that you have found in your investigation, because you will be able to think about it more easily when making a conclusion.

Scientists arrange data in ways that help them to spot patterns and trends. How they do this depends on the type of data.

How to display data

You can display data in three ways:

1. **Tables** – these set out data clearly so that you can see it all at once.

2. **Bar charts** – these show how something you are measuring changes, whether it's over time, with rising heat, with rising pressure, and so on. The time, heat or pressure goes along one axis (the *x*-axis), and the **variable** – the thing that changes – goes along the other axis (the *y*-axis).

3. **Line graphs** – these display data from investigations where you are looking at two variables, both of which change.

Graphs are really useful. When you look at them you can see a pattern in your results quickly. You can also see results that are out of place.

Tables

- *Do not forget to head table columns with quantity and unit.*
- *Make sure that your readings are reliable and recorded to the same degree of accuracy.*

Charts and graphs

- *Make sure graphs are as large as possible. Balance this with finding sensible scales for both axes.*
- *Make sure that you have placed the correct objects and quantities along the x- and y-axes.*
- *Look for a pattern in the points you plot on the graph.*

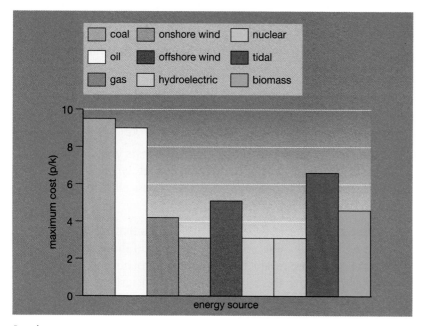

Bar chart

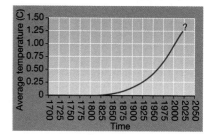

Line graph

When drawing a graph, always remember to label both axes and use the right units.

How well have you presented the data you have found?

Looking at results

- Look for results that don't look like they belong with the rest of the results. They could be extremely high points, extremely low points, or points with completely random locations. Try to work out why this is.

- Leave room for human error. An anomaly can be there because you made a mistake in measurement. You should ignore these results, as they won't tell you anything useful.

The units that your answer is in may not always be the same as the units that the question is in.

Thinking about what you've found

- If you've been asked to find a value, remember to include the unit that it's measured in.

- If you've made a graph, try to say something about its shape and what that indicates.

- If you can see a pattern on your graph, try to explain it in a short sentence.

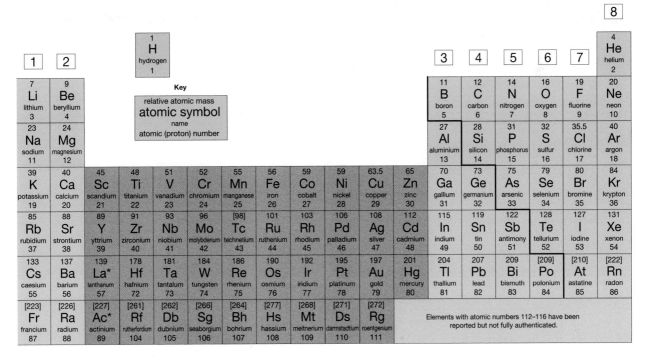

The periodic table

Evaluate your data

Scientists use data for **evidence**. It's like police work. Fingerprints, marks at the scene of the crime and DNA samples are all data that can be gathered. They are evidence that will later be used in court to get a conviction. Likewise, you will use your data to make a **conclusion**.

The big question

You should have asked yourself the following questions by now:

- Which areas of my data are the most reliable?

- Is there a pattern in the results I have got?

- Are there any unexpected results?

- Does this mean I have found something that answers my **big question**?

Your conclusion

If the answer to the last question is 'yes', you can make a conclusion. In your conclusion you evaluate the results you have produced.

To do this, you have to ask yourself:

- What pieces of evidence did I use to form my conclusion?

- Does the rest of my evidence support my conclusion?

- Were there any anomalies, and where?

- Why are they there? Was it human error, or something else?

- Is there any way I could have improved my data?

Conclusion or opinion?

Your **conclusion** must relate to the question asked: it is the answer to that question. You should explain why you have thought of this conclusion, and how sure you are that it is correct. You might not be so sure about it if your measurements are not very reliable, for example. Even if you're not sure about this, you can still have an opinion.

Remember your data

You should make your conclusion when thinking about the data you've collected and investigated. What has this told you?

You can (and should) talk about the data in your conclusion. This is how you back up your opinion. This is what makes it a **conclusion**, instead of just an **opinion**.

Remember the limitations of your data. An investigation conducted on members of your own school may not mean that your conclusions are true for the whole world!

Watch out

- You must base your conclusion entirely on the results you have. If you don't, it isn't a conclusion. It's just an opinion.

- There's no place for prejudice or hearsay when evaluating your results. Try not to let any outside factors influence your results, or what you think about them.

Health and Safety

Health and Safety is an important aspect of life today. This includes when we are working as scientists.

The number of injuries and deaths in science laboratories and workplaces where science is used is very small. But the risk is still present.

Dangers

Heat burns and scalds

Burns are caused by dry heat, like a Bunsen burner flame; scalds are caused by wet heat, such as boiling water.

The burned person will experience pain and possibly shock. Symptoms include swelling, blisters and redness.

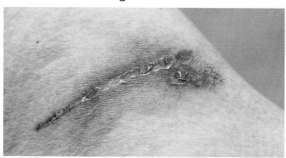

Chemical burns

Some chemicals can burn the skin if they come into contact with it. These may be liquids, such as a concentrated acid, or powders, such as lime.

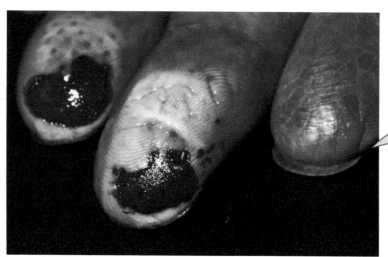

Sodium hydroxide burn

What to do

The priorities are to try to reduce the pain and the risk of infection. Cold water should be applied for several minutes and a dressing pressed to the affected area. If the burn is at all serious (the redder the skin, the more serious the burn), call 999.

Nothing should be applied to the injury other than water or a dressing.

What to do

The priorities are the remove the cause of the injury, and take steps to prevent infection. The cause of the burn should be removed, either with cool running water for at least 15 minutes in the case of a liquid, or brushing in the case of a powder. Any contaminated clothing or jewellery should be removed and the affected area wrapped loosely in a clean dressing or cloth.

Minor chemical burns usually heal themselves; however if there are symptoms of shock – or if the burn is deeper, you should seek expert attention.

What to do

The priorities are to ensure that the casualty's airway is open and to call an ambulance. If the casualty isn't responding, resuscitation may be necessary. If the chemical that has caused the injury is known, this should be communicated when making the call and to the ambulance crew as well.

What to do

Recovery cannot begin if the casualty is still in contact with the source of the current. In this case, additionally, the first aider is at risk of shock as well.

Removing the source of current may be done by switching off the supply using an insulator such as a wooden broom. If unconscious, a casualty's airway should be opened and resuscitation may be appropriate. An ambulance should be called.

Injury from breathing in fumes or swallowing chemicals

Many chemicals in scientific laboratories are potentially dangerous if swallowed. Some chemicals, such as ammonia, release fumes that are dangerous.

Symptoms vary according to what has been swallowed, but may include drowsiness, unconsciousness, nausea and vomiting, and even cardiac arrest.

Electric shock

Low voltages (typically around 12 V) present few risks in the laboratory. Mains voltage, however, can be lethal. If an electric current passes through the body, it can cause a person's breathing and heartbeat to stop.

Cuts and damage to the eyes from particles or chemicals

If foreign particles enter the eye, the priority is to remove them. It is important not to allow the casualty to rub his or her eyes as this will increase the damage or allow any other liquids to be introduced.

There are three established ways of removing particles. Let tears wash out the particle. Use an eye wash or Pull the upper lid down onto the lower lid and let the lower eyelashes sweep away the particles by blinking rapidly.

If this doesn't work, the eye should be closed and immediate expert attention sought. It is essential that the eye is not rubbed.

Hazard warning signs

These hazard warning signs are used very often in the laboratory. It is useful to learn what they mean off by heart, so that you can avoid harmful substances.

Corrosive

Irritant

Toxic

Highly flammable

Risk of electric shock

Oxidising

Glossary

absorb the small intestine absorbs food molecules – they pass from the small intestine through the thin walls into the blood stream

ABS braking system known as advance braking system which helps to control a skidding car

acid rain rain water which is made more acidic by pollutant gases

acupuncture treatment of pain or disease by inserting the tips of needles at specific points on the skin

accelerate an object accelerates if it speeds up

acceleration a measurement of how quickly the speed of a moving object changes (if speed is in m/s the acceleration is in m/s^2)

accuracy how close a reading is to its true value (forensic work needs accuracy)

acid phosphate test detects an enzyme present in semen

active site the place on an enzyme where the substrate molecule binds

addiction when a person becomes dependent on a medicine

addition polymer a very long molecule resulting from polymerisation, e.g. polythene

admissible admissible evidence is evidence which the trial judge finds useful in a trial

adrenal glands glands which make the hormone adrenaline and also corticosteroid hormones

adrenaline hormone which helps to prepare your body for action

adversarial justice system of justice where criminal trial courts operate with a prosecution team and a defence team

aerobic respiration respiration that involves oxygen

afterbirth the placenta is passed out through the vagina after birth and this is called the afterbirth

agglutination when red blood cells group together

agile being able to move in bursts in different directions

air bags cushions which inflate with gas to protect people in a vehicle accident

air resistance the force exerted by air to any object passing through it

alkalis substances which produce OH^- ions in water

alkali metals very reactive metals in group 1 of the periodic table, e.g. sodium

alkanes a family of hydrocarbons found in crude oil with single covalent bonds, e.g. methane

alkenes a family of hydrocarbons found in crude oil with double covalent bonds, e.g. ethene

allergic reaction excessive reaction of the body to a substance (pollen, dust or certain foods)

alloy mixture of two or more metals – used to make coins

alpha particles radioactive particles which are helium nuclei – helium atoms without the electrons (they have a positive charge)

alternating current an electric current that is not a one-way flow

alveoli small air sacs within the lung where gas exchange occurs

amino acids small molecules from which proteins are built

ammeter meter used in an electric circuit for measuring current

amniotic sac fluid-filled sac within the uterus – inside this the baby is protected from knocks

amps units used to measure electrical current

analgesic therapeutic drug used to reduce pain

anaemia disease where the body does not have enough red blood cells and therefore can't carry enough oxygen

anaerobic respiration respiration without using oxygen

anaesthetics drugs that cause temporary loss of body sensations

anagen phase actively growing phase of hair growth

analytical chemist someone who studies the chemical composition of materials

anion ion with a negative charge

anorexia nervosa eating disorder where the person eats less and less because they see themselves as being overweight (even though they may be underweight)

antenatal screening method for seeing if an unborn baby has abnormalities

anti-inflammatory drug used to reduce inflammation or swelling

antibacterial something which acts to kill bacteria, but is toxic to ingest e.g. antibacterial soap

antibiotic therapeutic drug acting to kill bacteria which is taken into the body

antibody protein normally present in the body or produced in response to an antigen which it neutralizes, thus producing an immune response

antidepressants drugs which help people suffering from depression

anti-diuretic hormone (ADH) hormone which controls re-absorption of water in kidneys (controls water levels in the blood)

antigen any substance that stimulates the production of antibodies – antigens on the surface of red blood cells determine blood group

antihistamine therapeutic drug used to reduce allergic reactions

anti-human serum blood serum which will give a precipitate if mixed with human blood

antistatic an antistatic particle cannot spark a fire in dangerous conditions

antiviral therapeutic drug acting to kill viruses

appendicitis inflammation of the appendix

aromatherapy the therapeutic use of aromatic plant extracts and essential oils in baths or massage

arteries blood vessels that carry blood away from the heart

arteriosclerosis disease of arteries where they become blocked

artificial active immunity where the body is given an inactive form of the pathogen to stimulate an immune response

artificial passive immunity where someone receives a readymade dose of antibodies against the infection

asteroids small objects in space which orbit a larger mass (e.g. the Earth)

atom the basic 'building block' of an element which cannot be chemically broken down

Atomic Emission Spectroscopy (AES) technique for analysing metal ions using a very hot plasma flame

atomic number the number of protons found in the nucleus of an atom

atria chambers of the heart that receive blood from the veins

autoimmune diesease disease caused by the immune system of the body attacking the body, e.g. type 1 diabetes

autonomic nervous system part of the nervous system that controls the body's internal environment

back-calculation calculation to decide on a person's blood alcohol level a few hours previously

bacteria single-celled micro-organisms which can either be free-living organisms or parasites (they sometimes invade the body and cause disease)

bacterial infection infection caused by bacteria invading the body

balanced diet eating foods (and drinking drinks) which will provide the body with the correct nutrients in the correct proportions

balanced equation chemical equation where the number of atoms on each side of the equation balance each other

balanced forces forces acting in opposite directions that are equal in size

bases solid alkalis

basic metabolic rate (BMR) amount of energy that someone needs just to keep their body processes tacking over

bauxite mineral which contains aluminium

beta particles particles given off by some radioactive materials (they have a negative charge)

bile chemical produced by the liver which digests fats

binder chemical which holds particles of pigment together in paint

biodegradable a biodegradable material can be broken down by micro-organisms

biodiesel fuel made from rape seed oil

biodiversity range of different living organisms in a habitat

bioethanol ethanol made from fermented corn or sugar cane

biofuels fuels made from plants – these can be burned in power stations

biological cause when a pathogen infects the body causing disease

biological diagnosis when a doctor runs tests to find out what is happening inside the patient's body

biological factors chemicals made by the body which influence our health, e.g. an imbalance of hormones; or organisms which cause disease, e.g. bacteria

biological indicators organisms which live in water – their presence or absence tells scientists how polluted water is

biological washing detergents detergents (used to wash clothes) which contain enzymes

biomass waste wood and other natural materials which are burned in power stations

biometric information information obtained by measuring and analysing biological data, e.g. fingerprints

bladder sac inside the body which acts as a temporary store for urine

blast furnace furnace for extracting iron from iron ore

blood fluid that is pumped through the body by the heart; contains plasma, blood cells and platelets

blood groups blood falls into one of four groups: A, B, AB or O

blood pressure force with which blood presses against the walls of the blood vessels

blood spatter pattern formed by blood as it sprays from the body

blood test test that investigates the composition of blood

blood transfusion when blood from one person is put into another person (the two people must have compatible blood groups)

body fluids liquids that are inside the bodies of living things, e.g. blood

body mass index (BMI) measure of someone's weight in relation to their height

bone marrow transplant bone marrow is taken from a healthy person and put into a person suffering from sickle cell anaemia – the 'new' bone marrow then makes normal red blood cells

braking distance distance travelled while a car is braking

breast screening method of checking for breast cancer at an early stage

breathing process which involves the exchange of gases between lung and air

buckyballs carbon molecules containing many carbon atoms which form large hollow spheres

budget an amount of money available for spending

buffer an aqueous solution of a mixture of a weak acid and a weak base (the pH of the buffer changes very little when a small amount of strong acid or base is added to it)

bulimia eating disorder where the sufferer alternates between eating too much and vomiting to keep to a chosen weight

calibration using known weight to draw up a scale of spring extension

calorific value the energy value in a food

cancer life-threatening condition where body cells divide uncontrollably

capillaries small blood vessels that join arteries to veins

carbohydrases enzymes which break down carbohydrate molecules

carbon-neutral fuel fuel grown from plants so that carbon dioxide is taken in as the plants are growing – this balances out the carbon dioxide released as the fuel is burnt

carboxylic acid an organic acid with a carboxyl group

carboxylic acid group the molecular group COOH which is found in organic acids

carcinogen something which causes cancer, e.g. chemicals in cigarettes

cardiac muscle muscle found in the heart which squeezes and relaxes continuously

cardiovascular fitness the fitness of the heart and circulatory system

carrier someone who carries an abnormal gene but does not themselves have the disease

case history detailed account of a patient's health and health history

cast positive image of an object made from an impression, e.g. a shoeprint is an impression which results in the cast of a shoe sole

catalyst substance added to a chemical reaction to alter the speed of the reaction

cation ion with a positive charge

cell membrane layer around a cell which helps to control substances entering and leaving the cell

central nervous system (CNS) collectively the brain and spinal cord

cervical screening method of detecting abnormal cells (cancerous cells) taken from the cervix

cervix opening of the uterus (the baby passes through the cervix and down the vagina)

chain of continuity when pieces of evidence from a scene of crime are passed from one investigator to another

chlorophyll pigment found in plants which is used in photosynthesis (gives green plants their colour)

CFCs gases which were used in refrigerators and which harm the ozone layer

chemical barrier substance in the body that can kill or limit the effect of a pathogen

chemical bond very strong force attracting together atoms in a molecule

chemical change any process in which one or more substances are changed into one or more different substances

chemical digestion enzymes carry out chemical digestion when they break down large food molecules into smaller ones

chemical energy energy found in fuels and foods

chemical evidence evidence relating to a scene of crime which is investigated by chemical analysis

chemical instrumental test test which involves gas chromatography or emission spectroscopy to analyse the purity of water

chemical property property which can't be observed just by looking at a substance – a chemical property depends on how that substance reacts chemically with other substances

chemical reaction a process in which one or more substances are changed into other substances – chemical reactions involve energy changes

chemical therapy type of treatment that uses drugs to treat an illness

chemotherapy drug treatment which kills or damages cancer cells

chiropractics system of treatment based on the idea that disease results from a lack of normal nerve function (chiropractors manipulate bones in the back to ensure they do not press on nerves)

chloride ion when a chlorine atom gains an electron it becomes a chloride ion with a negative charge – it is written as Cl⁻

cholesterol fatty substance which can block blood vessels

chromatogram the plate of results produced by thin-layer chromatography

chromatography process used for separating mixtures using differences in their absorbency

chromosomes thread-like structures in the cell nucleus that carry genetic information

cilia tiny hair-like structures which help to keep mucus and dust out of the lungs

circulatory system a transport system in the body that carries oxygen and food molecules

classify when we classify something we put it into a certain group

closed question question which only allows for a 'yes' or 'no' answer

coal fossil fuel formed from plant material

cognitive therapy treatment for depression which believes that changing thought patterns can solve problems

collision theory an idea that relates collisions among particles to their reaction rate

column chromatography technique where pigments are separated by passing through a column of silica gel

combustion process where fuels react with oxygen to produce heat

comets lumps of rock and ice found in space – some orbit the Sun

common approach path a single entrance to the crime scene and a path to the body

comparison microscope microscope that allows two images to be viewed at the same time

complementary therapy healing practices that do not fall within the realm of modern medicine, e.g. acupuncture

complete combustion when fuels burn in excess of oxygen to produce carbon dioxide and water only

components the parts of an electric circuit

compound two or more elements which are chemically joined together, e.g. H_2O

compressive force a force which squashes something

Computerised Tomography (CT) scan body scan which uses an x-ray beam that scans from many angles to build up a 3D picture

concave lens lens which is thinner in the middle than at the edges

conclusions opinions reached after studying all the evidence

conclusive test test which determines the presence, quality or truth of something

conduction heat transfer caused by particles in a solid vibrating and passing on thermal energy to nearby particles

conductors materials which transfer thermal energy easily; electrical conductors allow electricity to flow through them

conservation areas areas (can be on land or at sea) where there is special protection so that the wildlife is not disturbed

contaminants pollutants which affect land, sea or air, e.g. chemicals from factories

controlled conditions where conditions (e.g. temperature) are controlled during experiments

control sample uncontaminated sample which can be used for comparison (in forensic work)

convection heat transfer in a liquid or gas – when particles in a warmer region gain thermal energy and move into cooler regions carrying this energy with them

convection current when particles in a liquid or gas gain thermal energy from a warmer region and move into a cooler region, taking this energy with them

convex lens lens that is thicker in the middle than at the edges

cornea transparent outer covering to the eye

corrosive poisons poisons which 'gnaw away' at flesh, e.g. strong acids

cortex part of a hair which contains pigment granules (gives hair its colour)

corticosteroids hormones made by the adrenal glands which help the body cope with illness or injury

cosmic background radiation radiation coming very faintly from all directions in space

cost-benefit analysis an analysis of different alternatives to see whether the benefits outweigh the costs

cost-effective a cost-effective drug is one that works but at the same time is not extremely expensive

Court of Appeal a court to which appeals are made on points of law resulting from the judgment of a lower court

covalent bonds bonds between atoms where some of the electrons are shared

coverage the coverage of a screening programme means what percentage of the target population has been screened

Criminal Justice System (CJS) system responsible for detecting crime and bringing it to justice

critical angle angle at which a light ray incident on the inner surface of a transparent glass block just escapes from the glass

cross-contamination when evidence from a crime scene gets mixed up and contaminated with other materials

Crown Court court where more serious criminal cases are tried

Crown Prosecution Service a department of the government of the UK responsible for public prosecutions

crude oil black material mined from the Earth from which petrol and many other products are made

crumple zones areas of a car that absorb the energy of a crash to protect the centre part of the vehicle

crust surface layer of the Earth made of tectonic plates

current flow of electrons in an electric circuit

cuticle outer, protective layer of a hair

cystic fibrosis genetic condition where the lungs become clogged with mucus

cytology test test which looks at body cells and examines their appearance, e.g. cervical smear test

dative covalent bond type of covalent bond where both shared electrons come from the same atom

decelerates an object decelerates if it slows down

defendant person against whom a prosecution is brought in a court of law

defibrillator machine which gives the heart an electric shock to start it beating regularly

denatured an enzyme is denatured if its shape changes so that the substrate cannot fit into the active site

density the density of a substance is found by dividing its mass by its volume

dental records record of someone's teeth made by a dentist – everyone has a unique dental record

depression condition where people feel unhappy or worthless most of the time

diabetes disease where the body cannot control its blood sugar level

diagnosis identifying the nature or cause of a medical problem

diameter distance across a circle, from one side to the other and passing through the centre

diastolic blood pressure the lowest point that your blood pressure reaches as the heart relaxes between beats

diffuse when particles diffuse they spread out

digestion process by which your body breaks down large food molecules into smaller ones

digestive system the 9-metre long system that handles and digests food (starts at mouth, ends at anus)

diet what a person eats

diminished a diminished image is made smaller

direct current an electric current that flows in one direction only

displacement distance moved in a certain direction

displacement reaction chemical reaction where one element displaces or 'pushes out' another element from a compound

displayed formula when the formula of a chemical is written showing all the atoms and all the bonds

distribution living organisms are found in certain places – they have a distribution, e.g. some plant species are only found in warmer areas of the UK

dioxins polluting chemicals produced during the manufacture of PVC

distance–time graph a plot of the distance moved against the time taken for a journey

DNA molecule found in all body cells in the nucleus – it's sequence determines how our bodies are made (e.g. do we have straight or curly hair), and gives each one of us a unique genetic code

DNA fingerprint identification obtained by examining a person's unique sequence of DNA

DNA polymerase enzyme used to copy a DNA sample

DNA profile pattern of bands on an electrophoresis gel

documentary evidence evidence that includes statements from witnesses and CCTV footage

double covalent bond covalent bond where each atom shares two electrons with the other atom

drag energy losses caused by the continual pushing of an object against the air or a liquid

DVLA car registration database database record of all cars in the UK

dwarf planets objects orbiting the Sun which are smaller than planets and larger than asteroids

eating disorder abnormal eating habits that may involve taking in too little or too much food

earthquake shaking and vibration at the surface of the Earth resulting from underground movement or from volcanic activity

ecosystem a habitat and all the living things in it

efficacy a measure of how efficient something is

efficient a process in which losses are minimised

elastic potential energy energy gained and stored when an elastic object is stretched

electrical conductors materials which let electricity pass through them

electric current when electricity flows through a material we say that an electric current flows

electrical energy energy from the flow of electrons through a circuit (e.g. devices with plugs or batteries use electrical energy)

electrical insulators materials which do not let electricity through them

electrocardiogram (ECG) test that records the electrical activity and rhythm of the heart

electrolysis when an electric current is passed through a solution which conducts electricity

electromagnet a magnet which is magnetic only when a current is switched on

electromagnetic spectrum electromagnetic waves ordered according to wavelength and frequency – ranging from radio waves to gamma rays

electromagnetic waves a group of waves that carry different amounts of energy – they range from low frequency radio waves to high frequency gamma rays

electrons small particles within an atom that orbit the nucleus (they have a negative charge)

electrophoresis process where DNA pieces are put into a gel and a voltage applied across it separate the pieces according to size – this produces a DNA profile

electrostatic attraction attraction between opposite charges, e.g. between Na^+ and Cl^-

elements substances made out of only one type of atom

effectors cells that respond to a stimulus by moving the body, e.g. muscle cells

egg the female sex cell

emollient thick liquid with a soothing and moisturizing effect when applied to the skin

endocrine system the body system that is made up of endocrine glands (these secrete hormones)

endoscope device using optical fibres which allows doctors to look inside the human body

endoscopy technique for looking inside the body using a tiny camera on the end of an endoscope

endothermic reaction chemical reaction which takes in heat

energy the ability to 'do work' – the human body needs energy to function

energy-efficient choices choosing to use more energy-efficient equipment

energy profile diagram diagram showing energy taken in or given out during a chemical reaction

energy transfers when energy moves from one object to another or from one region to another

energy transformations when energy changes from one type to another

enzymes biological catalysts that increase the speed of a chemical reaction

epidemic when a disease spreads throughout a community

epithelium cells which line a surface, e.g. epithelial cells lining the lungs

esters chemical substances that give us fruit flavours

ethanol another name for pure alcohol

eutrophication when waterways become too rich with nutrients (from fertilisers) which allows algae to grow wildly and use up all the oxygen

evidence data resulting from scientific trials

evaporative cooling where water in sweat escapes from skin's surface and cools down the body

excretion the process of getting rid of waste from the body

exercise physical activity (running is a form of exercise); exercise can be **mild intensity**, **moderate intensity** or **high intensity**

exothermic reaction chemical reaction in which heat is given out

expert witnesses witnesses such as forensic scientists who provides opinions as well as factual information

expiration when air is forced out through the lungs

faeces the solid waste left at the end of the digestive process

faeces test medical test to see how well the digestive system is working and if there are any problems

fatty acids the building blocks of fats (together with glycerol)

feedback mechanism a mechanism involved in homeostasis which is switched on if the internal environment changes

fertilisation when a sperm fuses (joins with) an egg

fertiliser chemical put on soil to increase soil fertility and allow better growth of crop plants

fever when the temperature of the body rises above a 'normal' temperature of about 37°C

fibrillates the heart fibrillates if it twitches instead of beating

fingerprint patterns formed by skin ridges on fingers: can be **plastic**, **visible** or **latent**

fitness ability to do what your body requires you to do – includes speed and flexibility

flame test test where a chemical burns in a Bunsen flame with a characteristic colour – tests for metal ions

flammable a substance that burns easily

flavouring a substance that gives flavour to a food

flexible a person who has a wide range of movement around their joints

fluid any substance which flows, e.g. water and air

focal point the point where light rays refracted through a convex lens meet up

foetus the unborn baby within the uterus

follicle a ball of cells in the ovary containing an 'unripened' egg cell

follicle-stimulating hormone (FSH) the hormone in females (made in the pituitary) which stimulates a follicle in an ovary to develop into a mature egg

food additives substances, such as flavourings, added to food

food chain flow chart to show how a living thing gets its food

food web flow chart to show how a number of living things get their food (more complicated than a food chain)

forensic anthropology area of forensic science involving human skeletal remains

forensic entomologist forensic scientist who specialises in insects found on decomposing bodies

forensic investigation an investigation of evidence at a crime scene carried out by a forensic scientist

forensic odontologist forensic scientist specialising in the study of human teeth

Forensic Science Service (FSS) the provider of forensic advice to the Criminal Justice System in the UK

forensic toxicologist forensic scientist specialising in the effect of poisons and drugs

formulation the form that a medicine takes, e.g. tablet or liquid

fossil fuels fuels such as coal, oil and gas

fractional distillation column found in a refinery, this sorts the hydrocarbons in crude oil according to size of molecules

fractions the different substances collected during fractional distillation of crude oil

fragments very small pieces, e.g. fragments of glass

frequency the number of waves passing a set point per second

friction energy losses caused by two or more objects rubbing against each other

fullerenes cage-like carbon molecules containing many carbon atoms, e.g. buckyballs

fungal infection infection caused when the body is invaded by a pathogenic fungus

fungi living organisms which can break down complex organic substances (some are pathogens and harm the body)

fungicide chemical used to kill fungi

galaxy a group made of billions of stars

gametes the male and female sex cells (sperm and eggs)

gamma rays ionising electromagnetic waves that are radioactive and dangerous to human health – but useful in killing cancer cells

gas chromatography special type of chromatography which analyses the chemicals in paint – used in forensics

gene section of DNA that codes for a particular characteristic

gene therapy medical procedure where a virus is used to 'carry' a gene into the nucleus of a cell (this is a new treatment for genetic diseases)

general anaesthetic anaesthetic which puts the patient to sleep while surgery is carried out

genetic adaptation the way in which biological species adapt genetically to changing environmental conditions

genetic disorders inherited diseases passed on from parents to children

genetic investigations medical tests that check on DNA

genetic variation the differences between individuals (because we all have slight variations in our genes)

genome all the genes in an organism

genus a group consisting of more than one species (used in classification of living things)

geologist scientist who studies rocks and the formation of the Earth

geothermal energy energy obtained by tapping underground reservoirs of heat (energy from 'hot rocks')

giant ionic structure sodium chloride forms a lattice, also called a giant ionic structure

glands organs within the body which produce hormones

glomerulus a tiny ball-shaped clump of blood vessels that filters substances from blood (found in kidney tubules)

glucagon hormone produced by the pancreas which converts glucose stored in the liver to glucose

glucose a simple sugar (when combined with oxygen, glucose releases energy)

glycerol together with fatty acids, these make up fats

Gore-Tex™ synthetic fibre which keeps you dry inside and out

graphite a black solid formed from carbon atoms joined to make flat sheets

gravity an attractive force between objects (dependent on their mass)

greenhouse gas any of the gases whose absorption of solar radiation is responsible for the greenhouse effect, e.g. carbon dioxide, methane

group within the periodic table the vertical columns are called groups

Haber process industrial process for making ammonia

haemoglobin chemical found in red blood cells which carries oxygen

haemophilia disease where blood lacks the ability to clot

halide salt made from a group 7 element

halogens reactive non-metals in group 7 of the periodic table, e.g. chlorine

hazard something that is likely to cause harm, e.g. a radioactive substance

heart disease blockage of blood vessels that bring blood to the heart

heart rate the number of heartbeats every minute

herbicide chemical used to kill weeds

hertz units for measuring wave frequency

homeostasis the way the body keeps a constant internal environment

hormones chemicals that act on target organs in the body (hormones are made by the body in special glands)

host the organism on which a parasite lives (parasite takes food from the host)

Huntington's disease a genetic disease that causes writhing movements and usually strikes in middle age

hybrid cars cars powered by electric batteries which also have fuel engines

hydrocarbons molecules containing only carbon and hydrogen

hydroelectricity electricity generated by harnessing the power of moving water

hydrogen fuel cell a device which produces electricity from hydrogen

hydroxide ion consisting of oxygen and hydrogen atoms (written as OH^-)

hypertension where blood pressure is higher than the normal range – also known as **high blood pressure**

hypotension low blood pressure

identification key pathway used to help in identification of an organism

indicator chemical which changes colour according to the pH (indicators show how acid or alkali a substance is)

image an image is formed by light rays from an object that travels through a lens or are reflected by a mirror

immune system a body system which acts as a defence against pathogens, such as viruses and bacteria

immunity you have immunity if your immune system recognises a pathogen and fights it

immunological test a test which uses a function of the immune system, e.g. the precipitin test for blood which uses anti-human serum

incomplete combustion when fuel burns in a small amount of oxygen so that carbon monoxide and water are produced

indicator in electric equipment an indicator shows whether the device is switched on

inert an inert substance is one that is not chemically reactive

infection when the body has been invaded by pathogens (disease-causing organisms)

infertile if someone (male or female) is unable to have a child they are known as infertile

infectious if someone is infectious they can pass on a disease caused by pathogens to another person

infrared spectroscopy technique for analysing the chemical nature of a fibre

infrared waves non-ionising waves that give out heat – used in toasters and electric fires

ingest to take something into the body (we can ingest medicine)

inhaled medicine medicine breathed in through the nose

initial assessment evaluating the new situation – this is a police officer's first job when he or she gets to a scene of crime

injected medicine medicine inserted into the body using a needle

inorganic compound compounds made by non-living natural processes or in laboratories

impression mark made by some hard object causing a softer object to alter its shape

insoluble salt salt which is not soluble in water (forms a precipitate)

inspiration when air is taken into the lungs

insulator a material that transfers thermal energy only very slowly

insulin hormone made by the pancreas which controls the level of glucose in the blood

intermolecular force force between molecules

Intoxilizer machine for measuring alcohol in a person's breath

intravenous treatment where a drug is given directly into a vein

ionic compound in an ionic compound positive and negative charges are held together by **ionic bonds**, e.g. sodium chloride

ionic bond a chemical bond between two ions of opposite charges

ions charged particles (can be positive or negative)

ion chromatography process that allows the separation of ions and molecules based on the charge properties of the ions or molecules

iris surrounds the pupil in the eye and controls amount of light entering

iron core partly molten inner part of the Earth made up of molten and solid iron and other metals

isotope atoms with the same number of protons but different numbers of neutrons

judge public officer who hears and decides on legal cases and gives out justice

jury a body of citizens sworn to give a true verdict according to the evidence presented in a court of law

Kastle-Meyer (KM) test a forensic test for blood

keyhole surgery surgery carried out using an endoscope, which means that the doctor need only make small cuts on the body

kidney dialysis medical procedure for people whose kidneys don't work (dialysis filters and cleans their blood)

kidney tubules structures in kidneys which filter blood, removing certain substances from it

kinetic energy the energy that moving objects have

kingdoms classification of all living things into one of 5 groups – each group is called a kingdom

labour contraction of the uterus to push out the baby

lactic acid a substance produced from anaerobic respiration

lattice ordered structure formed by ions in an ionic compound, e.g. NaCl forms a lattice

law of reflection the angle of reflection is the same as the angle of incidence

laser therapy treatment using lasers, e.g. some eye surgery involves lasers

lava magma which has erupted onto the surface of the Earth

legal alcohol limit above a certain level of blood alcohol it is unsafe to drive (and illegal)

lens the convex lens found in the eye which focuses light

lenses pieces of glass or plastic shaped to bend and control the way light travels through them

leukaemia cancer of the white blood cells – it means that you are likely to catch many infections

life cycle the stages that some living things go through as they age, e.g. a fly goes through egg, larva, pupa and adult stages

lifestyle the way that someone chooses to live their life – this can be assessed in a lifestyle evaluation

light-dependent resistor (LDR) device in an electric circuit whose resistance falls as the light falling on it increases

light emitting diode (LED) a very small light in electric circuits that uses very little energy

light energy energy given off as light - anything that glows (such as a light bulb) gives out light energy

light microscope microscope that uses visible light and lenses to make small objects larger

limescale hard white substance found inside 'furred up' kettles (mostly calcium carbonate)

lime water calcium hydroxide particles in water – this clear liquid turns milky in the presence of carbon dioxide

lipases enzymes that break down fats into fatty acids and glycerol

litmus indicator that goes red with acids and blue with alkalis

LiveScan inkless electronic fingerprinting

local anaesthetic anaesthetic which numbs the area being operated on while the patient stays awake

long-sighted where you cannot see close objects clearly and the image forms behind the retina

long-term goals targets which may take some time to achieve

LPG fuel mixture of fuels such as propane and butane (short for liquefied petroleum gas)

luminal forensic chemical that enhances bloodstains

lung cancer when cells in the alveoli divide in an uncontrollable way and destroy lung tissue

luteinising hormone (LH) hormone (made in the pituitary) which, together with FSH, controls the release of an egg from the ovary

lymphocytes white blood cells that surround pathogens and make antibodies to destroy them

magistrate someone who administers the law in a **Magistrate's Court**

Magistrate's Court the lowest level of court in England and Wales where less serious cases are tried

magma molten rock found below the Earth's surface

Magnetic Resonance Imaging (MRI) scan body scan that uses a magnetic field and radio waves to create 2D or 3D images of the body on a computer screen

magnified an image is made larger

mammogram x-ray of breast to detect cancer

mantle semi-liquid layer of the Earth beneath the crust

Marquis reagent chemical used by forensic scientists to indicate the presence of some drugs

mass spectrometer instrument for identifying chemicals very accurately (chemicals are measured according to their mass)

mechanical digestion breaking food down into small pieces by non-chemical means

medulla air-filled channel in the centre of each hair

melanic form a form of moth which has dark colouration

melanoma dangerous type of skin cancer

menstrual cycle monthly hormonal cycle which starts at puberty in human females

menstruation monthly breakdown of the lining of the uterus leading to bleeding from the vagina

mental illness things going wrong with the mind

meteors bright flashes in the sky caused by rocks burning in Earth's atmosphere

microbial infection infection caused by micro-organisms

micro-organism very small organism (living thing) which can only be viewed through a microscope – also known as a microbe

microsurgery surgery carried out using very powerful magnifying devices

microtome cutting instrument which gives very thin sections

microwaves non-ionising waves used in satellite and mobile phone networks – also in microwave ovens

misdiagnosis when a patient is incorrectly diagnosed

mixtures one or more elements or compounds which have been mixed together – they can be separated out fairly easily

mobile phase in chromatography the water is the mobile phase

mobility ability to move

molecule two or more atoms which have been chemically combined

molecular formula the formula of a chemical using symbols in the periodic table, e.g. methane has a molecular formula of CH_4

monoculture when a single crop is grown in a huge field

monomer small molecule that may become chemically bonded to other monomers to form a polymer

moons large natural satellites which orbit a planet

motor neuron nerve cell carrying information from central nervous system to muscles

muscular dystrophy genetic disease where the muscles become increasingly weak

mutated a mutated cell is one where the genes have been altered

mutation where the DNA within cells have been altered (this happens in cancer)

myelin sheath insulating layer around a nerve fibre

nanochemistry the study of materials in the nanoscale range (1-10 nanometers)

nanometre units used to measure very small things (one billionth of a metre)

nanomotor very small motor to drive a nanoscale vehicle

nanoparticles very small particles on the nanoscale

nanoscale scale which deals in nanometres

nanosilver nanoparticle containing silver

nanotube carbon molecule in the form of a cylinder

narrow spectrum antibiotics antibiotics effective only against a few types of bacteria

National Grid network that carries electricity from power stations across the country (it uses cables, transformers and pylons)

national identity card identity card that each citizen in a country is issued with and required to carry

natural active immunity when the body 'remembers' a pathogen and launches an attack when it meets it again

natural fibres fibres from animal or plant sources

natural flavouring flavour that comes from natural materials

natural gas mixture of gases formed from animals and plants which lived 100 million years ago (it is a fossil fuel)

natural selection process by which 'good' characteristics that can be passed on in genes become more common in a population over many generations ('good' characteristics mean that the organism has an advantage which makes it more likely to survive)

nebulae clouds of dust and gases from which stars form

neonatal screening health checks carried out on newborn babies

nervous system body system for conducting messages (via nerves) around the body

neuron a nerve cell

neurotransmitter chemical that passes between nerve cells

neutral a neutral substance has a pH of 7

neutralisation reaction between H^+ ions and OH^- ions (acid and base react to makes a salt and water)

neutrons small particle which does not have a charge found in the nucleus of an atom

newtons unit of force (abbreviated to N)

NHS Primary Care Trust (PCT) part of the NHS responsible for the planning health services and improving the health of a local population

non-invasive a non-invasive surgical procedure does not involve making large cuts, e.g. keyhole surgery

non-rechargeable a non-rechargeable battery runs off a chemical reaction within the battery and cannot be topped up with mains electricity

non-renewable something which is used up at a faster rate than it can be replaced e.g. fossil fuels

normal a line at right angles to a boundary

nuclear energy energy that is stored inside an atom

nuclear fission energy released when nuclei of atoms are split

nuclear fuels radioactive fuels, such as uranium and plutonium, which are used in nuclear power stations

nuclear fusion energy released when two small atoms combine to make a larger atom

nuclear power energy which comes from slowly releasing the energy stored inside an atom (by nuclear fission)

nucleons protons and neutrons (both found in the nucleus)

nucleus central part of an atom that contains protons and neutrons

nutrients substances in food that we need to eat to stay healthy, e.g. protein

nutrition the way the body takes in and uses food

obesity a medical condition where the amount of body fat is so great that it harms health

object something which we view and from which light rays travel

oesophagus tube from the mouth that leads to the stomach

oestrogen female hormone secreted by the ovary and involved in the menstrual cycle

OH group an oxygen atom bonded to a hydrogen atom and found in all alcohols

ohms units used to measure resistance to the flow of electricity

oil fossil fuel formed from animals and plants that lived 100 million years ago

oil refinery an industrial plant that heats crude oil to separate it into its chemical components, which are then made into useful substances

open question the sort of question that invites a long and detailed answer

optic nerve nerve leading from the retina to the brain

optimum pH the pH at which an enzyme-controlled reaction works most effectively

oral medicine medicine given by mouth

ore rock which contains metal, e.g. iron ore

organ transplant where an organ from one person is transplanted into another person, e.g. liver transplant

organic acid an organic (naturally found) acid, e.g. vinegar

organic compound chemicals containing carbon, e.g. toxins found in animals and plants

organic solvent a liquid containing carbon that dissolves a solid, e.g. alcohol

osteopathy treating muscle or skeletal problems while also considering diet, lifestyle and posture (treatment is carried out by an osteopath)

osteoporosis disease of the bones in which bones become very brittle

ovaries organs (in females) which make eggs

overnutrition when a person eats too much

oviducts tubes which lead from ovaries to the uterus

ovulation release of an egg from the ovary

oxidising agent chemical that adds oxygen atoms

oxygen debt the debt for oxygen that builds up in the body when demand for oxygen is greater than supply

oxygenated blood blood containing oxyhaemoglobin

ozone gas found high in the atmosphere which absorbs UV rays from the Sun

pancreas organ which makes the hormones insulin and glucagon

pandemic when a disease spreads rapidly across many countries – perhaps the whole world

parallel circuit electric circuit formed by more than one loop so that the electrons can go through different paths

parasite organism which lives on (or inside) the body of another organism

pathogen harmful organism which invades the body and causes disease

pathology analysis of the results of biological tests, such as blood and urine tests

penis male organ which introduces sperm into the female's vagina

peripheral nervous system network of nerves leading to and from the brain and spinal cord

periodic table a table of all the chemical elements based on their atomic number

peristalsis muscle action which moves food through the digestive system

petrochemical industry industry concerned with crude oil and the materials produced from it

pesticide chemical used to kill living organisms which are pests, e.g. rats or insects

pH meter a device which measures the pH of a substance accurately

pH scale scale running from 0 to 14 which shows how acid or alkali a substance is

phagocytes white blood cells that surround pathogens and digest them with enzymes

pharmacist a person trained to carry out the instructions on a doctor's prescription and administer drugs

photochromic photochromic materials react to the stimulus of light

photosynthesis process carried out by green plants where sunlight, carbon dioxide and water are used to produce glucose and oxygen

physical barrier something that stops pathogens entering the body, e.g. the skin

physical cause when an illness is caused by an injury

physical change change from one state (solid, liquid or gas) to another

physical diagnosis when a doctor observes physical symptoms such as a rash

physical factor something in the environment that affects health, e.g. living in damp, cold conditions

physical illness things that can go wrong with the body – either infectious disease or non-infectious

physical property property that can be measured without changing the chemical composition of a substance, e.g. hardness

physical therapy treatment that does something physically to the body

physiological effects something which influences the way the body works, e.g. exercise has a physiological effect when it strengthens heart muscle

physiological investigation taking measurements about a person's level of fitness, e.g. pulse rate

physiotherapy exercises and massage to treat muscle problems

piezoelectric a piezoelectric material produces a voltage when under stress

pigment material which gives paint its colour

pituitary gland gland at the base of the neck which makes many different hormones

placenta organ that develops in the female during pregnancy through which the foetus obtains oxygen and food molecules

planet large ball of gas or rock travelling around a star (planets orbit our Sun and Earth is an example)

plaque build up of cholesterol in a blood vessel (which may block it)

plasma yellow liquid found in blood

plastic compounds produced by polymerisation, capable of being moulded into various shapes or drawn into filaments and used as textile fibres

plasticisers small molecules which fit between polymer chains and allow them to slide over each other

platelets cell fragments which help in blood clotting

pollution contaminating or destroying the environment as a result of human activities

pollutant gases gases released into the air which damage the environment (gases in car exhausts)

poly(ethene) also called polythene, this useful plastic is made from ethene gas

polymer large molecule made up of chains of monomers

polymerisation chemical process that combines monomers to form a polymer: this is how polythene is formed

polymerase chain reaction (PCR) process whereby many copies of DNA can be made

positive charge when an atom loses an electron it become an ion with a positive charge, e.g. Na^+

potential difference another word for voltage (a measure of the energy carried by the electric current)

potential energy another word for stored energy

powdering forensic technique for collecting latent fingerprints on hard surfaces

power source source of electricity (a battery or mains electricity)

precipitate solid formed in a solution during a chemical reaction

precipitin test a conclusive forensic test used to test for human blood

precipitation reaction chemical test in which a solid precipitate is formed – tests for metal ions

precision how close a series of repeated results are to each other (forensic work needs precision)

predator animal which preys on (and eats) another animal

prescription a doctor's written orders for the preparation and administration of a drug

presumptive test a positive result for a presumptive test means that the scientist can presume that the result is positive, but can't be absolutely sure (this test can give false positive results)

prey animals which is eaten by a predator

product molecules produced at the end of a chemical reaction

professional ethics code of behaviour used by forensic scientists – they have to present information in an impartial way

professional witnesses people such as police officers who provide factual evidence in court

progesterone hormone produced by the ovary which prepares the uterus for pregnancy

proteases enzymes which break down proteins into amino acids

proteins molecules made up of amino acids – we need proteins in our diet (found in food of animal origin and also in plants)

protons small positive particles found in the nucleus of an atom

psoriasis painful skin condition where the skin cracks

psychological effect something which affects the way the mind works, e.g. regular exercise can help to fight off depression

psychotherapy treatment of psychological problems by allowing the patient to talk these through with a trained therapist

physiological to do with the way the body works

pulse rate the movement of blood in your arteries (taking your pulse rate shows how fast your heart is beating)

pupil dark hole in the middle to the eye that allows light to enter

pyrogram print-out given by gas chromatography

PVC a polymer (short for polyvinylchloride)

PVCu unplasticised polyvinylchloride

qualitative analysis an analysis of a chemical (or chemicals) to find out what's in it

quantitative analysis an analysis of a chemical (or chemicals) to find out how much is present

radiation thermal energy transfer which occurs when something hotter than its surroundings radiates heat from its surface

radiologists people trained and experienced in interpreting body scans

radiotherapy using ionizing radiation to kill cancer cells in the body

radio waves non-ionising waves used to broadcast radio and TV programmes

rate of reaction the speed with which a chemical reaction takes place

ray diagram diagrams showing how light rays travel

reactants chemicals which are reacting together in a chemical reaction

reaction force when an object feels a force it pushes back with an equal reaction force in an opposite direction

reaction time the time it takes for a driver to step on the brake after seeing an obstacle

real evidence exhibits from the crime scene

real image image formed on the other side of the lens to the object

receiver device which receives waves, e.g. a mobile phone

receptors nerve cells which detect a stimulus, e.g. a hot surface

rechargeable a battery is rechargeable if it can be recharged with a flow of mains electricity

recovery time how long it takes for the pulse and breathing rate to return to normal after exercising

red blood cells blood cells which are adapted to carry oxygen

red shift when lines in a spectrum are redder than expected – if an object has a red shift it is moving away from us

reflecting telescope telescope that reflects and focuses light using a concave mirror

reflex an muscular action that we take without thinking about

reflex arc pathway taken by nerve impulse from receptor, through nervous system, to effector (does not go through brain)

refraction when a light ray travelling through air enters a glass block and changes direction

refracting telescope telescope that uses two convex lenses to collect and focus light

refractive index amount by which a piece of glass will bend a ray of light

rehabilition the restoration of someone to a useful place in society

renal system body system that produces, collects and gets rid of urine

renewable energy that can be replenished at the same rate that it's used up e.g. biofuels

representative sample sample from a scene of crime which can be used for different kinds of tests

replacement therapy a physical therapy that involves replacing parts of the body that are not working well

replicating when organisms or cells make more copies of themselves

resistance measurement of how hard it is for an electric current to flow through a material

resistant when bacteria are resistant to an antibiotic they are not killed by it (the antibiotic fails to work)

respiration process occurring in living things where oxygen is used to release the energy in foods

respiratory system body system involved in respiration (in humans these are the lungs and trachea)

resting pulse rate normal pulse rate when not exercising

resultant force the combined effect of forces acting on an object

retina covering of light-sensitive cells at the back of the eyeball

reversible a reversible change (or reaction) is one that can also work in the opposite direction

rhesus blood group system blood can be grouped into rhesus-positive and rhesus-negative groups

Rf value Rf stands for retention factor (Rf value indicates distance moved by a colour divided by the distance used by water) – used in chromatography

ridge patterns patterns made by ridges of skin found on the fingers – loops, whorls and arches

risk the likelihood of a hazard causing harm

risk assessment deciding on the level of risk in a particular course of action

saliva clear liquid secreted into the mouth by the salivary glands which moistens the mouth

saturated fat fats, most often of animal origin, which are solid at room temperature

sat-navs Navigation system using location and traffic information from orbiting satellites

scanning electron microscopy (SEM) using a microscope that magnifies by scanning the sample with high-energy electrons

Scene of Crime Officer (SOCO) police officer who specialises in dealing with crime scenes

screening test medical test where large numbers of people are tested for a disease before the symptoms have appeared also called screening programmes

scrotum pouch of skin containing the testis

sedative substance that lowers a person's level of awareness

seismograph instrument for measuring and recording the vibrations of earthquakes

sequence arrangement in which things follow a pattern

semen fluid that contains sperm

semiconductor substance that does not conduct electricity at low temperatures but does so at higher temperatures

sensor device that detects a change in the environment

sensory neuron nerve cell carrying information from receptors to central nervous system

series circuit circuit formed by a single loop of electrical conductors

shells electrons are arranged in shells (or orbits) around the nucleus of an atom

short-sighted where you cannot see distant objects clearly and the image forms in front of the retina

short-term goals targets which can be achieved fairly quickly

sickle cell anaemia disease caused by sickle-shaped red blood cells which cannot carry enough oxygen around the body

side-effects unwanted effects produced by medicines

single-celled organism a living organism made up of one cell only, e.g. a bacterium

single covalent bond bond between hydrogen atoms where each atom shares its electron with the other

skin graft when a layer of skin that has been damaged is replaced by another skin layer from elsewhere in the body

smart materials materials which change with a stimulus

sodium ion when the sodium atom loses an electron it becomes an ion with a positive charge, Na^+

sodium hydroxide test test for metal cations – the colour of the precipitate formed indentifies the cation

soil erosion where soil is worn away (for example by wind or heavy rain)

solar cells devices which convert the Sun's energy into electricity

solar energy energy from the Sun

solar panels panels which use the Sun's energy to heat water

soluble salt salt which dissolves in water

solutes substances which dissolve in a liquid to form a solution

solution when a solute dissolves in a solvent a solution forms

solvents liquids in which solutes dissolve to form a solution

solvent front distance moved by water in chromatography

solvent system mixture of solvents used in chromatography

somatic nervous system the part of the nervous system concerned with voluntary actions

sound energy anything making a noise gives out sound energy

sound waves vibrations which can travel through solids, liquids and gases – we detect them with our ears

species basic category of biological classification, composed of individuals that resemble one another, can breed among themselves, but cannot breed with members of another species

specimen a sample animal or plant taken for investigation

spectrum pattern of light given off when a chemical burns (analysed by a spectroscope)

speed how fast an object travels: speed = distance ÷ time

sperm the male sex cell

sperm ducts tubes which carry sperm

sphygmomanometer blood pressure monitor

spring constant a number indicating how stretchy the spring is

star bright object in the sky which is lit by energy from nuclear reactions

state symbols symbols used in equations to show whether something is solid, liquid, gas or in solution in water

stationary phase in chromatography the paper used is the stationary phase

statistics branch of mathematics concerned with the collection and interpretation of quantitative information

stem cells unspecialised body cells (found in bone marrow) that can develop into other, specialised cells that the body needs, e.g. blood cells

step down transformer a step-down transformer changes alternating current to a lower voltage

step up transformer a step-up transformer changes alternating current to a higher voltage

stethoscope medical instrument for listening to sounds generated within the body, e.g. heartbeats

stimulus something that stimulates receptor nerve cells, e.g. a hot surface

stroke sudden change in blood flow to the brain – can be fatal

stomach acid acid produced by the stomach (helps in digestion)

subcutaneous a subcutaneous injection is given just under the skin

subscript small number in a chemical formula, e.g. in H_2O the small numeral 2 is the subscript

sub-station where transformers change the voltage in electrical cables

substrate molecules at the start of a chemical reaction

successful collisions when particles with enough kinetic energy react with each other

surgery operating on the body to explore, remove or repair structures

symptom a visible sign in a patient or a change in the way their body feels

systolic blood pressure the highest point that your blood pressure reaches as the heart beats to pump blood through your body

synapse gap between two neurons

synthetic fibres fibres made by man-made processes

tape-lifting forensic technique used to collect fibres

target organ the part of the body affected by a hormone

telogen phase phase of hair growth where the hairs have stopped growing

tensile force a force which stretches something

terminal velocity the top speed reached when drag matches the driving force

testes organs (in males) where sperm are made

test print forensic technique for capturing the imprint of a shoe

tetrahedral structure very strong three-dimensional structure (carbon atoms can be linked like this)

thalassaemia inherited disease where damaged blood proteins are made – causing anaemia

therapeutic drug substance used in the diagnosis, treatment or prevention of an illness

thermal conductors materials which allow heat to pass through them easily, e.g. metals

thermal energy another name for heat energy

thermal insulators materials which do not conduct thermal energy easily, e.g. wood

thermite process exothermic reaction between iron oxide and aluminium which results in molten iron

thermistor sensor in an electric circuit that detects temperature

thermochromic a thermochromic material is one that changes with heat

thermoregulation how the body keeps a constant internal temperature

thin-layer chromatography (TLC) chromatography using a glass sheet coated with powder instead of paper

thinking distance distance travelled while the driver reacts before braking

thyroid gland gland at the base of the neck which makes the hormone thyroxin

thyroxin hormone made in the thyroid gland which controls metabolic rate

titration common laboratory method of used to determine the unknown concentration of a known reactant.

tissues groups of cells that work together and carry out a similar task, e.g. lung tissue

topical a topical medicine is one that is rubbed onto the skin

total internal reflection gland at the base of the neck which makes the hormone thyroxin

toxic a toxic substance is one which is poisonous, e.g. toxic waste

toxin poisonous substance (pathogens make toxins which make us feel ill)

trace evidence evidence from crime scenes present in very small amounts

transfer stain stain produced when blood is transferred to a clean surface

transformer device by which alternating current of one voltage is changed to another voltage

translation the assembling of amino acids to build a protein

transmitter device which transmits waves, e.g. a mobile phone mast

transparent a transparent substance can be easily seen through, e.g. clear glass

transplant when an organ is taken from one person and surgically placed into another person, e.g. liver transplant

triple covalent bond bond between nitrogen atoms where each atom shares three electrons with the other

tsunami huge waves caused by earthquakes – can be very destructive

tumour abnormal mass of tissue that is often cancerous

turbine device for generating electricity – the turbine moves through a magnetic field and electricity is generated

ultrasound high-pitched sounds which are too high for detection by human ears

ultraviolet radiation electromagnetic waves given out by the Sun which damage human skin

umbilical cord structure that links the foetus to the placenta

undernutrition when a person does not consume enough calories or nutrients

unsaturated fats fats which are liquid at room temperature and come from plants or fish

urea nitrogen-containing substance cleared from the blood by kidneys and excreted in urine

urinary system body system that produces, collects and gets rid of urine

urine liquid waste from kidneys which is excreted from the body

urine test test to investigate what the body is excreting through the kidneys

uterus organ where a baby develops

universal indicator indicator which shows the pH of a substance by changing colour (useful for the entire range of the pH scale)

universe the whole of space

vaccine killed micro-organisms, or living but weakened micro-organisms, that are given to produce immunity to a particular disease

vagina also called the birth canal, this leads from the uterus (when a baby is born it passes down the vagina)

vascular screening health checks to find out the risk of heart disease

vasoconstriction in cold conditions the diameter of small blood vessels near the surface of the body decreases – this reduces the flow of blood

vasodilation in hot conditions the diameter of small blood vessels near the surface of the body increases – this increases the flow of blood

valves structures in the heart that prevent backflow of blood

vector a carrier that transfers DNA (genes) – modern medicine uses viruses as vectors to carry genes

veins blood vessels that carry blood back to the heart

velocity how fast an object is travelling in a certain direction: velocity = displacement ÷ time

velocity–time graph a plot of the velocity of a moving object at all times during its journey

ventricles chambers of the heart that pump blood into the arteries

villi 'finger-like' structures on the surface of the small intestine which give it a greater surface area for absorption

viral infection infection caused when the body is invaded by viruses

virtual image image formed on the same side of the lens as the object; a virtual image formed by reflection can be seen but cannot be projected onto a screen

virtual reconstructions virtual reality solutions for crime scene investigation

viruses very small infectious organisms that reproduce within the cells of living organisms and often cause disease

viscosity a measure of how easily a liquid flows

visible light waves in the electromagnetic spectrum that we can detect with our eyes

volatile a volatile substance evaporates easily

volcano a landform (often a mountain) where molten rock erupts onto the surface of the planet

voltage a measure of the energy carried by an electric current (also called the **potential difference**)

volts units used to measure voltage

wavelength distance between two wave peaks

wave speed how quickly a wave carries energy from one place to another

well – insulated a well – insulated building contains materials designed to minimise heat loss

white blood cells blood cells which defend against disease

wide spectrum antibiotics antibiotics god at fighting a wide range of bacteria

wind energy energy contained in moving air which spins a turbine to generate electricity

witness evidence evidence where witnesses give factual information about what they saw or heard

x-rays ionising electromagnetic waves used in **x-rays photography** (where x-rays are used to generate pictures of bones)

yeast single-celled fungus used in making beer and bread

yield useful product made from a chemical reaction

Acknowledgements

The publishers wish to thank the following for permission to reproduce photographs. Every effort has been made to trace copyright holders and to obtain their permission for the use of copyright material. The publishers will gladly receive any information enabling them to rectify any error or omission at the first opportunity. (T = Top, B = Bottom, C = Centre, L = Left, R = Right):

Science Photo Library, p11TL/Lawrence Migdale, p14C/Andrew Lambert Photography, p20B/Charles D. Winters, p21C/Andrew Lambert Photography, p22C/Andrew Lambert Photography, p26T/Charles D. Winters, p27CL/Charles D. Winters, p39TC/NASA, p62TC/Antonia Reeve, p70BR/Detlev Van Ravenswaay, p201BL, p215BL/Martin, p242BR/Mauro Fermariello, p249L/Mauro Fermariello, p252CR/Volker Steger, Peter Arnold Inc., p259TR/Jim Varney, p269BL/Michael Donne, University of Manchester, p272R/Astrid & Hanns-Frieder Michler, p276C/Volker Steger, Peter Arnold Inc.&BR/Eye of Science, p280BL/Andrew Lambert Photography, p303TL/National Cancer Institute, p328B/Mehau Kulyk, p332BL/Andrew Lambert Photography, p333TL/Andrew Lambert Photography, p334BL/Charles D. Winters, p336BR/Andrew Lambert Photography, p341C/Andrew Lambert Photography, p348TR/Geoff Tompkinson, p350C/Sinclair Stammers, p362BR/Dr P. Marazzi

NI Syndication, 198T, 224T, 232T, 246T, 321BL

p17C/Liz Hoffman, p32TC/ Green Chemistry Centre of Excellence, based at the University of York. The Centre aims to promote the implementation of green and sustainable chemistry into new products and processes (www.greenchemistry.net), p52CR, totalcycling.com, p80CR/National Human Genome Research Institute, p82T/Reptile & Amphibian Ecology International, Paul S. Hamilton/RAEI.org, p112TC/Photograph reproduced with kind permission of De Beers UK Ltd., p118CR/Courtesy of Canland UK (Hot Pack) Ltd. producers and distributors of Hot Pack Self Heating Nutritious Meals. TM and Non Magnesium Flameless Ration Heaters, p122B/Schuyler S, p127TR/Ben Mills and Ephemeronium, p149B/GNU licence, p154/Display bookshelf reproduced with kind permission of Daniel Eatock 2009, p160R/NASA, p169C/Mr Fergus Paterson, Knee Surgeon at the Bupa Cromwell Hospital, SW5 0TU, p203BL/Centers for Disease Control and Prevention's Public Health Image Library (PHIL), p228T/Alex Wild, p233BL/Almazi, p245BL/Forensic Pathways Ltd www.forensic-pathways.com, p259B/Phadebas®, Magle AB Life Sciences, p258TC/UWA Mark Reynolds, p263TL&CL/Stuart James, p305CL/Graham Colm, p315CL/Sjbrown, p319TL/www.talktofrank.com (Home Office, Department of Health and the Department of Children Schools and Families), p324C/GNU Licence, p339BL/Slower, p349CL/Ching-Wan Yip & Dr. Yue-Ling Wong

iStockphoto, p8BL, p8-9, p13BC, p14TR, p14CR&BR, p15CL&R, p16BR, p18TC, p20TR, p24T&B, p25T, p30B/Stahlkocher, p 32BC, p33TL, p34TC, p35TL, p36TC&BR, p38TC, p42TC, p43BL, p48TC, p49TL&BR, p50T&CR, p51TL, p52T, p53TL&CL, p54CR, p55CL, p56TR&CR, p57CR, p59BL, p60CR, p61CL, p63BR, p64TL, p65BR, p68TC, p69BC, p70TC, p71CL, p77T, p76-77, p78BR, p82BC, p83BR, p91BR, p92TL/Tom Walker, p102TC, p105TR, p110B, p111T, p110-111, p118BR, p121CL, p126CR, p129BL, p138TC, p146B, p147TR, p146-147, p148TC&CR, p150TC&CR, p152BR, p153TL&BR, p154TC, p155BR, p159CL, p162TC&CR, p164CR, p165TL, p166CR, p168TC, p169CL, p173TL&BR, p174TL, p177TL&CL, p180TL&BR, p181BR, p180-181, p183TL, p187TL&CR, p189TL&BL, p194T, p196T, p197BR, p203C, p208TR, p209BC, p208-209, p210TC, p217BL, p218TL, p223BR, p236TC, p237BL, p243TL, p242-243C, p242&243, p244TL, p245BL, p246BR, p247TL, p248T&BR, p250BR, p251TL, p254TC, p255CR, p256CR&BR, p258C, p264TC&CR, p265CR, p266C, p268TC, p269CL, p270CR, p274TC, p278TC, p280TC, p282TC, p283CR, p287CL, p292TR&BC, p293TR, p294C, p298C, p300CR, p305CL, p306C, p312C, p314C, p316TC, p316BL, p319CL, p308TL, p308CR, p309TL, p309CB, p329CR, p328-329, p334CT, p340TR, p343L, p350TC, p352TC, p353BL, p356C, p360C, p360BL, p361C

Shutterstock, p9CL, p10TC, p11BC, p12CL, p15TL, p16T, p17TC, p19BL, p25BC, p28T&CR, p37C, p41CL, p43TL, p46TC, p47CR, p46-47, p48R, p50BR, p51CR, p55TL, p57CL, p58TC, p64BR, p76B, p78TC, p80C, p82BL,CR,CB, B&BR, p83BL,BL,C,BC,R&BR, p84TC, p85L, p86TC, p87TL,L&CL, p88T&B, p89TL,BL&C, p90BL&BR, p92CR, p93L, p94T, p95TL, p96TC, p98TC, p100TC, p104BR, p106T, p113TL, p115C, p116TC, p117C&CL, p120TC, p121L, p122TC, p123TL,C&BL, p124TC, p125C, p127TC&TL, p128TR, p129C, p130T&B, p131T&B, p132T, p133TL, p134T&R, p135TL,CR&BR, p137L, p138B, p139TL&BR, p140T&BR, p141BR, p142T, p143B, p151CL, p152T p154R, p156T, p158T&B, p159B, p171BL, p172T, p175BL, p176T&B, p182T, p183B, p184T, p185TL, p186T, p187BL, p188T&BR, p190T&BR, p191BL, p192C, p193C, p194C&BR, p198BR, p199T, p201TL, p202BR, p204T, p209TC, p216T, p219L, p220T, p225L, p230C&B, p237BL, p254BR, p258T, p260B, p265TL, p269TL, p270T, p272T, p274BR, p276T, p286T, p292-293, p295TL&CR, p296B&BR, p297C&BL, p298T, p300T&R, p301BL, p302T, p303CL, p304BR, p305TL, p306R, p307L, p308T&R, p309TL, p310CR, p311C&CL, p313TL&C, p315TL, p317C, p318T, p320T, p322T, p323TL&BL, p325BL, p329CL, p330T, p332T, p335TR, p337TL, p338T&R, p339TL&TR, p340C, p342T, p343R, p344T, p347T, p349B